ACS SYMPOSIUM SERIES **336**

Pesticides
Minimizing the Risks

Nancy N. Ragsdale, EDITOR
U.S. Department of Agriculture

Ronald J. Kuhr, EDITOR
North Carolina State University

Developed from a symposium sponsored
by the Division of Agrochemicals
at the 191st Meeting
of the American Chemical Society,
New York, New York,
April 13–18, 1986

American Chemical Society, Washington, DC 1987

Library of Congress Cataloging-in-Publication Data
Pesticides: minimizing the risks.
(ACS symposium series, ISSN 0097-6156; 336)
Includes bibliographies and indexes.
1. Pesticides—Congresses. 2. Pesticides—Safety
measures—Congresses. 3. Pesticides—Toxicology—
Congresses.
I. Ragsdale, Nancy N., 1938-   . II. Kuhr, Ronald J.
III. American Chemical Society. Division of
Agrochemicals. IV. American Chemical Society.
Meeting (191st: 1986: New York, N.Y.) V. Series.

SB950.93.P476   1987   632′.95   87-1842
ISBN 0-8412-1022-5

ACS Symposium Series

M. Joan Comstock, *Series Editor*

1987 Advisory Board

Foreword

The ACS SYMPOSIUM SERIES was founded in 1974 to provide a medium for publishing symposia quickly in book form. The format of the Series parallels that of the continuing ADVANCES IN CHEMISTRY SERIES except that, in order to save time, the papers are not typeset but are reproduced as they are submitted by the authors in camera-ready form. Papers are reviewed under the supervision of the Editors with the assistance of the Series Advisory Board and are selected to maintain the integrity of the symposia; however, verbatim reproductions of previously published papers are not accepted. Both reviews and reports of research are acceptable, because symposia may embrace both types of presentation.

Contents

Preface..vii

1. Minimizing the Risk Associated with Pesticide Use: An Overview..........1
   Alvin L. Young

TOXICOLOGY

2. Current Toxicology Requirements for Registration.....................14
   Raymond A. Cardona

3. Acute Versus Chronic Toxicity and Toxicological Interactions Involving
   Pesticides...20
   Raymond S. H. Yang

4. New Approaches for the Use of Short-Term Genotoxicity Tests
   To Evaluate Mutagenic and Carcinogenic Potential....................37
   Frederick J. de Serres

5. Simulation Modeling in Toxicology..................................43
   J. T. Stevens and D. D. Sumner

PESTS

6. Vulnerability of Pests: Study and Exploitation for Safer Chemical
   Control..54
   Robert M. Hollingworth

7. Pests as Part of the Ecosystem.....................................77
   L. V. Madden

CHEMICALS

8. Principles Governing Environmental Mobility and Fate................88
   James N. Seiber

9. Mammalian Metabolism..106
   H. Wyman Dorough

10. Molecular Modeling: A Tool for Designing Crop Protection Chemicals....115
    Erich R. Vorpagel

POTENTIAL HAZARD

11. Pesticide Use: The Need for Proper Protection, Application, and
    Disposal..128
    W. K. Hock

12. Educating the Public Concerning Risks Associated with Toxic
    Substances..139
    Ronald W. Hart and Angelo Turturro

13. Mass Media's Effect on Public Perceptions of Pesticide Risk:
    Understanding Media and Improving Science Sources.................149
    JoAnn Myer Valenti

v

PANEL DISCUSSION

14. Summary and Discussion..**168**
     Robert E. Menzer

Author Index...**175**

Affiliation Index...**175**

Subject Index..**175**

Preface

PUBLIC CONCERN HAS GROWN STEADILY over the past three decades concerning the effects of pest control chemicals on human health and environmental quality. A substantial legislative base currently exists for regulation of these chemicals to protect the environment and promote human safety. The regulations that result from this legislative mandate are based on risk assessment and risk management. Risk is defined as the potential adverse health effects resulting from exposure to environmental hazards. Regulations to reduce risk influence the integration of agricultural research and technology into society and stimulate new technological developments in efforts to resolve the unanswered questions associated with the risk assessment process.

The National Academy of Sciences indicated in a recent study that the primary problem with risk assessment is the sparse data upon which decisions are based. Agriculture has an opportunity to meet the need for improved risk assessment and management through new research, new technology, and the use of that technology.

The symposium upon which this book is based focused on the new directions we must follow, based on our current level of understanding. The speakers identified research and educational opportunities that will strengthen the data base that is used to minimize risk while maintaining the necessary quality and quantity of food and fiber.

We would like to express our appreciation to the U.S. Department of Agriculture for its interest in and support of this symposium. We would also like to thank Herbert Cole, Richard Honeycutt, and Christopher Wilkinson for their excellent ideas in the planning phase.

NANCY N. RAGSDALE
U.S. Department of Agriculture
Washington, DC 20251

RONALD J. KUHR
North Carolina State University
Raleigh, NC 27695

December 1986

Chapter 1

Minimizing the Risk Associated with Pesticide Use: An Overview

Alvin L. Young

Office of Science and Technology Policy, Executive Office of the President, Washington, DC 20506

Although the benefits of pesticides are undeniable, attention in recent years has been focused on their impact on human health and environment. Although pesticide law requires that both risks and benefits be considered in all decisions, risk drives the process in terms of depth of analysis and allocation of federal resources. Two questions relative to risk are appropriate: "What is acceptable risk?" and "How can we minimize the risk?". No amount of research can eliminate all uncertainties associated with assessing the risks of exposure to pesticides or eliminate the controversial judgments inherent in any decision about control of pesticide exposures. Yet, there are opportunities to reduce the scientific uncertainties and to increase the public's confidence that health is protected and the economic consequences of imposed pesticide control are justified.

The decades of the 1940's and 1950's were characterized by major discoveries in the use of chemicals that aided the development of a highly successful agricultural system, promoted the economic strength of the nation, and secured the public's health from the dread of vector-transmitted diseases. In the early 1960's, tremendous progress was made in application technology for the dissemination of pesticides. Development in innovative equipment for applications accompanied major advances in formulation chemistry and the widespread efficacy testing of pesticide products. The late 1960's and 1970's were characterized by the concerns for the environment and the implementation of a massive regulatory program "that now governs all environmentally relevant aspects of our economy" (1).

Today, pesticides continue to play a major role in our society; yet, an unprecedented level of controversy over their use continues. The focus of the controversy is in the determination of "what is acceptable risk". The answer to this question cannot be provided by science - it is a social question. What can be addressed by the scientific community is the question of "how can we minimize the risk."

Table I. Magnitude of Pesticide Use in the United States

Land Use Category	All Pesticides		Herbicides		Insecticides		Fungicides	
	Treated Hectares (x10⁶)	Quantity (x10⁶ kg)	Treated Hectares (x10⁶)	Quantity (x10⁶ kg)	Treated Hectares (x10⁶)	Quantity (x10⁶ kg)	Treated Hectares (x10⁶)	Quantity (x10⁶ kg)
Agricultural Lands	114	341	109	199	34	74	10	68
Government and Industrial Lands	28	55	30	44	--	11	--	--
Forest lands	2	4	2	3	1	1	--	--
Household Lands	4	55	3	26	3	25	1	4
TOTAL	148	455	144*	272	38	111	11	72

*Same land may be treated with several classes of chemicals.

SOURCE: Modified from Pimentel and Levitan (4), 1986.

This book, and the American Chemical Society's Symposium on which it
is based, addresses this latter question.

Extent of Pesticide Use

A generally acceptable definition of pesticides includes key phrases
such as "chemical substance or mixture of substances intended for
preventing, destroying, repelling, or mitigating any pests" and
"substances intended for use as a plant regulator, defoliant or de-
siccant (2,3). In 1980, some 530 million kg of pesticides were used
in the production of food, clothing and durable goods for the more
than 270 million persons living in the United States, that is 2 kg
per person. Most recently, Pimentel and Levitan (4) have estimated
that the current annual use of pesticides in the United States approaches about 500 million kg, primarily synthetic organic chemicals
but also including 27 million kg of sulfur and copper sulfate fungicides. Table I shows the use by major land-use categories for
herbicides, insecticides and fungicides. In 1983, the sites (users)
of pesticide applications in the United States were agricultural
lands (68%), industrial and commercial sites (17%), home and gardens
(8%) and government lands (7%)(5). The use in the amounts of herbicides and insecticides has dramatically changed in the past two
decades. Table II gives the amount of herbicides and insecticides
for selected years, applied annually by U.S. farmers. The growth in
herbicide use has been reflected in materials used in wheat, corn
and soybean, e.g. 2,4-D, atrazine, and trifluralin, respectively.
(Table III). As Ware (3) pointed out, pesticides are big business.
The United States market is the world's largest, representing 34%
of the total. In 1980, U.S. manufacturers produced 660 million kg
of synthetic organic pesticides, valued at $4.2 billion and the
retail value of pesticide sales in the United States reached $5.8
billion. Clearly, for such extensive demand and use there must be
significant benefits associated with the use of pesticides. Con-
versely, such quantities used annually means that misapplications,
accidents and improper use of pesticides constitute real, and docu-
mented risks associated with pesticide use.

Table II. Total Pesticides Used by U.S. Farmers on Crops

Year	Herbicides (Million Kilograms)	Insecticides (Million Kilograms)
1964	34.5	64.9
1966	50.8	62.6
1971	101.6	71.7
1976	178.7	73.5
1982	196.4	26.8

SOURCE: Council on Environmental Quality (1)

Table III.   Selected Herbicides Used by U.S. Farmers on Crops, 1982

Herbicide	Kilograms Used (Million)	Hectares Treated (Million)
Atrazine	34	21
Alachlor	38	18
2,4-D	10	17
Trifluralin	16	17
All Other	99	73
TOTAL	197	146

SOURCE: Council on Environmental Quality (1)

Benefits of Pesticide Use

Pesticides are an integral and indispensable part of American (and world) agriculture. Hayes (5), Mellor and Adams (2), Pimental and Levitan (4), and Ware (14) have discussed the benefits derived from the use of pesticides. Ware (3) noted that the plants that supply the world's main source of food are susceptible to 80,000 to 100,000 diseases caused by viruses, bacteria, mycoplasma-like organisms, rickettsias, fungi, algae, and parasitic higher plants. They compete with 30,000 species of weeds the world over, of which approximately 1,800 species cause serious economic losses. Some 3,000 species of nematodes attack crop plants, and more than 1,000 of these cause severe damage. Among the 800,000 species of insects, about 10,000 plant-eating species add to the devastating loss of crops throughout the world. Pimentel and Levitan (4) estimate that "total worldwide food losses from pests amount to about 45% (of total food production). Preharvest losses from insects, plant pathogens, and weeds amount to about 30%. Additional postharvest losses from microorganisms, insects, and rodents range from 10 to 15%.

In the United States crop losses due to pests are about 30% or $20 billion annually, despite the use of pesticides and other current control methods (3). Ware (3) addresses the question of "what would the losses be without the use of insecticides?" Studies conducted in 1976-1978 compared the yields from test plots, where insecticides were used to control insects, to adjacent plots in which the insects were allowed to feed and multiply uncontrolled. Summary data from this study are shown in Table IV. These data suggested that, on the average, for the crops evaluated, half of the crops are lost to insects. Even with insecticides, 10% of the crop is lost. The issue, however, is what rate of return can a farmer expect from the use of insecticides on these crops? The data suggested an average increased yield of 36%. Pimentel and Levitan (4) concluded that in economic terms, for the $3 billion invested in the United States in controlling pests through the use of pesticides, about $12 billion are returned in increased return on investment.

Table IV.   Comparison of Losses Caused by Insects in Plots Treated
            by Conventional Use of Insecticides and Untreated Plots

| Commodity | Calculated Losses (%) | | |
	With Treatment	Without Treatment	Increased Yield (Percentage)
Corn	17.7	42.2	24.5
Soybeans	5.5	20.8	15.3
Wheat	9.5	65.0	52.0
Cotton	14.5	51.1	39.1
Potatoes	1.0	48.0	47.0
Average of Above Crops	9.6%	45.4%	36.0%

SOURCE: Summarized from data presented by Ware (3)

Hayes (5) noted that although information is available on the
benefits of pesticides to agricultural production, information is unfortunately fragmentary on the benefits of pesticides to the protection of stored products. Losses that occur during storage are caused
mainly by insects, mites, rodents and fungi, all of which are susceptible to chemical control. It has been estimated that at least 10%
of harvested crops worldwide - enough to feed 200 million people
without additional land or cultivation - are lost during storage.
In the tropics, the loss is mainly in the quantity of food stored.
In temperate climates, the loss is mainly of quality and acceptability (5).

Hayes (5) has also reviewed the contribution of pesticides to
the control of human diseases spread by arthropods and other vectors.
Outbreaks of malaria, louse-borne typhus, plague, and urban yellow
fever, four of the most important epidemic diseases of history, have
been controlled by use of the organochlorine insecticides, especially
DDT. In fact, the single most significant benefit from pesticides
has been the protection from malaria. Today malaria eradication is
an accomplished fact for 619 million people who live in areas once
malarious. Where eradication has been achieved it has stood the
test of time. An additional 334 million people live in areas where
transmission of the parasite is no longer a major problem. Thus,
about 1 billion people, or approximately one-fourth of the population
of the world, no longer live under the threat of malaria.

In summary, pesticides have been and continue to be an integral
part of American public health and agricultural programs. The benefits of the proper use of pesticides are enormous. Nevertheless,
there are significant risks associated with widespread and intensive
use of pesticides.

Risks Associated with Pesticide Use

Much attention in recent years has been focused on the social and
environmental risks associated with pesticide use. Mellor and Adams
(2) conclude that human poisonings are clearly the highest price paid
for using pesticides. They reported that in many developing coun-
tries, improper use of pesticides by untrained workers has often
led to poisonings during application. Also, workers who entered
treated areas too soon after treatment to weed or harvest crops have
been exposed to pesticides by brushing against contaminated foliage.
Mellor and Adams estimated that in Central America a total of about
3,000 to 4,000 pesticide poisonings occur annually. Pimentel and
Levitan (4) estimate that 45,000 total human poisonings occur annually worldwide, including about 3,000 cases admitted to hospitals and
200 fatalities, with approximately 50 of the latter being attributed
to accidental death.

     Mellor and Adam (2) pointed out that by disrupting natural con-
trols, pesticides can have a detrimental impact on the environment.
For example, the use of the fungicide benomyl to control fungi in
soybeans may unleash damaging outbreaks of foliage-feeding caterpil-
lars that the fungi might otherwise have destroyed. When outbreaks
of new pests occur, additional control treatments must be used.
About $153 million is spent each year to control these newly created
pests (4).

     Pimentel and Levitan (4) estimated that less than 0.1% of the
pesticides applied to crops actually reaches the target pests.
Hence, they concluded that most of what is applied enters the environment, contaminating the soil, water, and air and perhaps poisoning
or adversely affecting nontarget organisms. The figure of 0.1%,
however, is calculated as the amount of insecticide, for example,
that comes in direct contact with the insect. The bulk of the pesticide is certainly within the environment of the target pests. In
the case of a herbicide, a far higher percent of the chemical inter-
sects with the target plant. The point is that we have much to
learn in how to direct the chemical to the target pest.

     To comprehend the magnitude of environmental pollution by pesti-
cides, Sun (6) recently reported on studies by the Environmental
Protection Agency showing that 17 pesticides have now been detected
in the ground water of 23 states; the concentrations typically ranged
from trace amounts to several hundred parts per billion. Two years
ago (1984) the count was 12 pesticides found in 18 states; but Agency
scientists attributed the rise to an increase in the quality and
quantity of studies rather than an increase in the problem. The
goal of the EPA surveys is to obtain sufficient information to
characterize pesticide contamination and to determine how pesticide
concentration levels correlate with patterns of usage and groundwater vulnerability factors. While the studies are not designed to
estimate individuals' pesticide exposure nor the resulting level of
health risk, the data will allow inferences about populations potentially at risk.

Brattsten et al. (7) have reported on another problem associated with the widespread use of pesticides; namely, the rapid
appearance of insecticide resistance. By 1980, 260 species of
agricultural arthropod pests had insecticide-resistant strains,
compared to 68 for disease vectors. They noted that some insects
have developed resistance to all major classes of insecticides
and will develop resistance to future insecticides as long as
present application techniques and use patterns prevail.

In summary, the current widespread use of pesticides have
resulted in problems involving human health, adverse effects on
nontarget organisms, pesticide resistance in many major pests, and
environmental contamination of air, soil and water. Many of these
problems, however, result from the improper use, handling, or storage
of pesticides. The recognition that pesticides pose risks to health
and the environment is the salient reason for government and industry
to undertake programs to minimize these risks.

Research Approaches to Minimizing Risks Associated with Pesticide Use

No amount of research can eliminate all the uncertainties associated
with assessing the risks of exposure to potentially hazardous chemicals. No amount of research can eliminate the controversial
judgments inherent in any decision about the control of chemical
exposures. Yet, there are opportunities to reduce scientific uncertainties and to increase the public's confidence that health and
environment are protected and that the economic consequences of
imposed chemical control are justified. The following research
areas have been identified for minimizing pesticide risks:

1. Minimizing Risk Through a Better Understanding of Toxicology

    ° Toxicology requirements
    ° Acute versus chronic toxicity testing
    ° New methods for toxicologic evaluation
    ° Simulating modeling

2. Minimizing Risk Through a Better Understanding of Pests

    ° Physiological and biochemical considerations
    ° Pests as part of the ecosystem

3. Minimizing Risks Through a Better Understanding of Potential
    Hazard

    ° Proper protection and disposal
    ° Educating the public
    ° Perceptions through the media

The remainder of this book will focus on each of the above
areas. Many of these areas have been the target of research teams
and have already shown potential payoffs. In 1983, Nelson (8) identified five toxicology research needs; namely 1) improve the basis
for dose and interspecies extrapolation to humans; 2) predict effects

of multiple chemical exposure; 3) develop cellular and molecular
markers of exposure; 4) develop means to distinguish carcinogens
on the basis of modes of action; and 5) expand use of existing federal data collection activities. As a consequence of Nelson's
report, the Federal Government has committed major resources to the
investigations of these areas. The key to minimizing risk is knowledge. The more we know and understand about the toxicology, mode
of action and environmental fate of a pesticide, the more "management conscious" we can be about the use of that pesticide. Accordingly, in late 1983 the National Pesticide Information Retrieval
System (NPIRS) was established. NPIRS is a computer data base that
describes key characteristics of 50,000 pesticide products registered by EPA plus thousands of state registrations. This data base
is currently updated weekly. NPIRS is maintained at Purdue University in Indiana and is supported through cooperative agreements with
the USDA and EPA.

     Research by Brattsten et al. (7) have stressed the need for
managing the problem of insecticide resistance. They suggested that
studies of the basic biology of insect-plant interactions in nature
and in crop agroecosystems can produce ideas for improved use of
chemicals and how they can best be integrated with nonchemical
methods. Gebhardt et al. (9) have emphasized the need to devise improved pest management strategies for conservation tillage and to
better understand the impact of conservation tillage on water quality, especially as it is related to use of agricultural chemicals.
Boyer (10) has recognized that research into understanding the basic
factors that influence plant productivity will significantly enhance
our effectiveness of using plant protection chemicals.

     Menn (11) has addressed the minimizing of risks through a better
understanding of chemical structure-activity relations. He discussed strategies leading to the discovery of selective and biodegradable
insecticides and insectostatic agents. Pimentel and Levitan (4) and
Eue (12) have identified the need to understand chemical structure
and the relationship to environmental mobility, bioaccumulation and
persistance. Eue (12) also addresses the need for standardization of
methods for studying the behavior of fate of herbicides in soil.

     In 1985, Benbrook (13) challenged the scientific community to
assist in the development of a national policy on pesticide residues.
He identified two critical areas for scientific input: 1) expertise
in developing the analytical and scientific insights needed to reform
our existing patchwork of laws and programs, and 2) sound data needed
for conducting reliable risk assessments to support regulatory decision making. He concluded his paper by stating: "Toxicologists need
to clearly articulate the benefits to society that could result from
more reliable up-to-date information on pesticide use patterns, and
hence exposure. Such knowledge would make it possible to begin dif-
ferentiating beween significant potential hazards that deserve regulatory scrutiny, and truly insignificant, improbable risks that
simply do not warrant the same degree of attention."

In assessing the regulation of pesticides, John Young (14), Chairman of the President's Commission on Industrial Competitiveness, stressed the need to balance environmental, health, and safety regulation with the needs of research, development and technological innovation. He concluded: "As the 21st Century approaches, Americans need to weigh more heavily the importance of maintaining a 'technology gap' in our favor if the Nation is to remain competitive in world markets and retain one of the highest standards of living in the world. Americans cannot (and do not) expect both absolute safety and a prominent position on the cutting edge of science and technology. Fear of the unknown and the unwillingness to accept small risks could make Government regulation an enemy of innovation. This could occur because a fundamental problem in our regulatory process is the failure to uniformly and properly balance safety concerns with the needs for innovation and industrial competitiveness."

Funding of Pesticide Research

The previous section identified numerous research areas. The funding and scientific staffing (in scientist years) of such research must be the responsibility of the entire scientific enterprise to include state and federal government and private industry. Data in Tables V and VI show FY 84 or CY 84 investments in pesticide research by government and industry, respectively. Table VII provides data for the federal expenditures for toxicology-related research.

Table V. Estimated FY 84 Federal and State Expenditures for Pesticide Research

Area	Funding	SYS
Fundamental Biology	$ 86.4 M	618
Improved Means of Nonpesticidal Control	74.5 M	498
Improved Pesticide Use Patterns	33.8 M	232
Toxicology, Pathology, Metabolism and Fate of Pesticides	9.6 M	62
Economics of Pest Control	3.9 M	26
TOTAL	$208.2 M	1,436

SOURCE: USDA Cooperative State Research Service

Table VI. CY 1984 Industrial Expenditures for Pesticide Research

Areas	Total Dollars	Total SYS
Synthesis and Screening	$ 124 M	776
Product Development & Registration	308 M	1,925
TOTAL	$ 432 M	2,701

SOURCE: Agricultural Research Institute (15)

Table VII.   FY 85 Federal Expenditures for Toxicology-Related
              Research

          $ 194 M - Basic Research (e.g. Mode of Action)
          $ 149 M - Toxicology Testing
          $ 56 M - Methods Development

     TOTAL $ 399 M

              88 Scientists Years

SOURCE: National Institutes of Health and National Science
           Foundation

     In our society, the issue of risk from pesticide use is one
where the level of public information is low and the potential for
anxiety high -- a situation likely to confound the policy-making
process. We do not have the luxury of discarding the "chemical
tools" necessary for control of the many pests that threaten our
food supply and health. Therefore, the task for agricultural scientists and policy makers is to effectively communicate risks and
benefits impartially to the public. Only through such a process can
we hope to retain the public's confidence in our abilities to safely
use pesticides. To that end, universities, industries and governments must commit resources to the conduct of essential pesticide
research; research that will meet the needs of the Nation's agricultural, industrial and public health programs.

Literature Cited

1.   Council on Environmental Quality. 1986. Environmental Quality.
      15th Annual Report of the Council on Environmental Quality.
      Superintendent of Documents, Washington, DC. 719 pp.
2.   Mellor, J.W. and R.H. Adams, Jr. 1984. Feeding the Under-
      developed World. Special Report, Chem & Eng. News, April 23,
      1984. P 32-39.
3.   Ware, G.W. 1983. Pesticides: Chemical Tools. Chapter 1,
      P. 3-25. IN: Pesticides: Theory and Application. H.W. Freeman
      and Co., New York, NY.
4.   Pimentel, D. and L. Levitan. 1986. Pesticides: Amounts Applied
      and Amounts Reaching Pests. BioScience 36:86-91.
5.   Hayes, W.J., Jr., 1981. Toxicology of Pesticides. The
      Williams & Wilkins Co., Baltimore, MD.
6.   Sun, M. 1986. Ground Water Ills: Many Diagnoses, Few
      Remedies Science 232:1490-1493.
7.   Brattsten, L.B., C.W. Holyoke, J.R. Leeper, and K.F. Raffa.
      1986. Insecticide Resistance: Challenge to Pest Management
      and Basic Research. Science 231:1255-1260.
8.   Nelson, N. (Chairman). 1983. Research Briefing Panel on
      Human Health Effects of Hazardous Chemical Exposures. Com-
      mittee on Science, Engineering and Public Policy. National
      Academy Press, Washington, DC P 100-110.

9.   Gebhardt, M.R., T.C. Daniel, E.E. Schweizer, and R.R. Allmaras.
     1985. Conservation Tillage. Science 230:625-630.
10. Boyer, J.S. 1982. Plant Productivity and Environment.
     Science 218:443-448.
11. Menn, J.J. 1983. Present Insecticides and Approaches to
     Discovery of Environmentally Acceptable Chemicals for Pest
     Management. IN: Natural Products for Innovative Pest Manage-
     ment. Whitehead, D.L. and W.S. Bowers (Eds). Pergamon Press,
     New York, NY P 5-31.
12. Eue, L. 1985. World Challenges in Weed Science. Weed Science
     34:155-160.
13. Benbrook, C.M. 1985. National Policy Review. Panel on Pesti-
     cide Residues: Policy, Practices and Protection. Board on
     Agriculture, NAS/NRC. Presentation to the Toxicolocy Forum,
     July 17, 1985. Aspen, CO. 5 pp.
14. Young, J.A. (Chairman). 1984. Balancing Environmental, Health,
     and Safety Regulation with the Needs of Research, Development,
     and Technological Innovation. Appendix C, Vol. 2:279-293.
     Report of the President' Commmission on Industrial Competitive-
     ness. Government Printing Office, Washington, DC.
15. Agricultural Research Institute. 1985. A Survey of U.S.
     Agricultural Research by Private Industry III. ARI, Bethesda,
     MD., 26 pp.

RECEIVED September 25, 1986

TOXICOLOGY

Chapter 2

Current Toxicology Requirements for Registration

Raymond A. Cardona

Uniroyal Chemical Company, Inc., Crop Protection Research & Development, Bethany, CT 06525

The Environmental Protection Agency (EPA) is responsible for pesticide regulation and for insuring the safety of pesticide products. To this end, the EPA has promulgated testing guidelines designed to fulfill pesticide registration requirements. Laboratory animal studies form the primary basis for predicting the potential hazards of pesticides to public health. EPA toxicology data requirements include acute testing, subchronic and chronic feeding/oncogenicity studies, a two generation reproduction and teratogenicity study, mutagenicity testing and a rodent metabolism study. These requirements and problems encountered in interpretation of data obtained from toxicity testing are discussed. Future directions of research to fulfill toxicology data requirements are described.

The Environmental Protection Agency (EPA), under the authority of the Federal Insecticide, Fungicide and Rodenticide Act (FIFRA), is responsible for pesticide regulation. Data requirements for pesticide registration, and guidelines for evaluating the toxicity of pesticides to nonhuman organisms and for relating the results of these studies to human safety evaluations, have been developed by EPA(1,2).

Toxicity studies required by EPA for pesticide registration are listed in Table 1. The major toxicity categories include acute, subchronic, chronic and mutagenicity testing. Evaluation of teratogenicity and adverse reproductive effects are included under chronic testing. Animal metabolism and dermal penetration studies are found in a special testing category.

Acute Toxicity

The initial step in the safety evaluation of a pesticide product is the determination of its acute toxicity (Table I). Laboratory animals, usually rats and rabbits, are exposed to a single dose of the test substance. Toxic effects resulting from ingestion,

0097-6156/87/0336-0014$06.00/0

inhalation, skin and eye contact are determined over a two to three
week post-exposure observation period. If repeated dermal exposure
is expected to occur, a dermal sensitization study in guinea pigs
is required. An acute delayed neurotoxicity study is required only
if the test substance is for food use and is an organophosphate,
which causes acetyl cholinesterase depression, or if it is structurally related to a substance that causes delayed neurotoxicity. The
purpose of acute toxicity testing is to establish relative toxicity
to other chemicals by defining the median lethal dose (LD50 and
LC50), to provide initial information on the mode of toxic action,
to identify possible synergistic interactions and to evaluate design
for subchronic tests. Data from these tests are used to determine
the signal word and the hazard warning statements which appear on
the pesticide product label; for example, if the pesticide is
acutely toxic (oral LD50 in rat <50 mg/kg) the signal word Danger
must be on the label.

Table I.   EPA Toxicology Data Requirements

TEST	SPECIES
ACUTE	
°Oral/Dermal/Inhalation	Rat, Rabbit
°Primary Eye/Dermal Irritation	Rabbit
°Dermal Sensitization	Guinea Pig
°Delayed Neurotoxicity	Hen
SUBCHRONIC	
°90 Day Feeding	Rat and Dog
°90 Day Dermal/Inhalation	Rat
°90 Day Neurotoxicity	Hen
CHRONIC	
°Oncogenicity	Rat and Mouse
°Chronic Feeding	Rat and Dog
°Teratogenicity	Rat and Rabbit
°Reproduction, 2-Generation	Rat
MUTAGENICITY	
°Gene Mutation	
°Chromosome Aberration	
°DNA Damage and Repair	
SPECIAL	
°Metabolism	Rat
°Dermal Penetration	Rat

Subchronic Tests

Following the determination of its acute toxicity, the pesticide is
then evaluated in subchronic studies (Table 1). Subchronic exposure
usually lasts for 90 days and its objective is to generally evaluate
and characterize the effects of the test substance when administered
to laboratory animals on a daily basis. EPA requires that subchronic studies be performed in two species, the rat and dog, and at
several dosage levels. The test substance is normally administered
to the animals in their diet. However, when pesticide use involves
purposeful application to or prolonged exposure of human skin, or if

its use results in repeated inhalation exposure at a concentration
likely to be toxic, 90 day studies by the dermal and/or inhalation
route will be required. If acute results show signs of neurotoxicity, a 90 day study is needed to further evaluate this end point.

Parameters which are evaluated in the subchronic testing phase
include: general observation of test animals for clinical signs of
toxicity, body weight changes, diet consumption, mortality, organ
weight changes, clinical chemistry measurements, gross necropsy and
histopathology. In addition to providing information on target
organs and possible test substance tissue accumulation, the results
serve to estimate the dosage levels to be used in long term or
chronic toxicity tests.

Chronic Tests

Requirements under the category of chronic testing are listed in
Table 1. Adverse effects resulting from long-term exposure to a
pesticide are evaluated. Oncogenicity studies assess the potential
of the test agent to produce malignant and benign tumors and pre-
neoplastic lesions. They are performed in the rat and mouse and
extend over the majority of the expected life span of the strain,
about 2 years for rats and 18 months for mice. Chronic feeding
studies, carried out in the rat and a non-rodent species, usually
the dog, are designed to evaluate other chronic effects in addition
to tumor formation. EPA believes that testing in a non-rodent
species is necessary to provide an adequate evaluation of non-
oncogenic effects. Chronic and subchronic feeding studies are
similar except the period of exposure to the test substance is
longer in chronic tests, i.e. 2 years duration for rats and 1 year
duration for dogs.
    EPA includes two tests in the category of chronic evaluation
which are designed to examine the effects of a pesticide on the
process of reproduction. A teratogenicity study is performed in two
species, the rat and rabbit, to evaluate potential fetotoxicity or
birth defects in offspring. Pregnant animals are exposed to the
test substance daily by gavage during that part of pregnancy
covering the critical period of organogenesis, or the time during
the development or growth of organs, since teratogens are most
effective at this stage of gestation. Organogenesis corresponds to
days 6 to 15 of gestation for the rat and days 6 to 18 of gestation
for the rabbit. In humans, it is complete during the first tri-
mester of pregnancy. Parturition, the process of giving birth,
occurs on day 21 and 32 for the rat and rabbit, respectively.
Fetuses are delivered by cesarean section one day prior to parturition and examined for gross, visceral and skeletal abnormalities. A
two generation reproduction study is conducted in the rat. The test
substance is administered to male and female rats in their diet
prior to mating. Treatment of inseminated females is continued
during their growth into adulthood, mating and production of the
second generation. Effects of a test substance on gonadal function,
estrus cycles, mating behavior, conception, parturition, lactation,
weaning, and the growth and development of the offspring are
measured.
    In general, chronic studies must incorporate several, usually

three, dosage levels. The EPA requires that the highest dose used
in these studies must be the maximum tolerated dose (MTD) or one
that produces some toxic or pharmacological effect in the test
animal. Also, a lower dose level must be used which produces no
evidence of toxicity. This is the "no observed effect level" or
NOEL. The use of the MTD in chronic assays can cause problems in
interpretation of test results especially when the MTD is responsible for overloading of metabolic pathways. Overloading enzymatic
transformation of a test substance and/or metabolite results in
disproportionate changes in blood and tissue levels and relative
quantities of parent substance and metabolites. An expanding number
of studies show that metabolic pathways can be overloaded, sometimes
at doses well below a toxic level or MTD. When such overloading
occurs in a toxicity study, the true hazard to another species cannot be reliably estimated (3). To avoid overloading of metabolic
pathways, pharmacokinetic and metabolism experiments should be conducted prior to the start of chronic studies. Results from these
experiments in conjunction with results from subchronic studies,
should be used for determining the highest dose level used in
chronic tests. In this way a better understanding of the actual
hazard of the pesticide will be obtained since the chronic study
will be carried out at dosage levels which will not alter the animals normal physiological response to the test substance.

Other Tests

Other toxicology requirements needed prior to registration include a
battery of mutagenicity studies and a rodent metabolism and dermal
penetration study (Table 1). The rodent metabolism and dermal penetration protocols are included under the category of special testing.
Mutagenicity data is required to determine if the pesticide will
affect genetic components in the nucleus of the mammalian cell. If
a mutation is present in the genetic material of the egg or sperm at
the time of fertilization, the resulting consequences may be embry-
onic or fetal death or congenital abnormalities. The battery of
mutagenicity assays includes tests to detect gene mutations, structural chromosome aberrations and other genetic effects such as DNA
damage and repair and mammalian cell transformation. These mutagenic assays are also used to screen for potential carcinogens since
the initial step in chemical carcinogenesis is generally believed to
be a mutagenic event. A rat metabolism study is required to determine the transformation, absorption, distribution and excretion of
the pesticide. Dermal absorption evaluations are needed for pesticides having a serious toxic effect, as identified by oral or inhalation studies, or for which a significant route of human exposure is
dermal penetration.

The main purpose of the toxicity tests just described is to
provide a data base that can be used to evaluate the hazard and
assess the risk associated with the use of a pesticide. In practice,
the no observable effect level (NOEL) found in the most sensitive
animal species tested in chronic studies is used. To extrapolate a
safe dose for human consumption, a safety factor of 100 is usually
used. For example, if the NOEL in the most sensitive animal species,
e.g. the dog from the chronic feeding study, was 10 mg/kg of body
weight, then the acceptable daily intake (ADI) for man would be

0.10 mg/kg of body weight. Since the NOEL is by definition a sub-
threshold dosage level, this safety factor approach would not be
applicable to pesticides that are carcinogenic and mutagenic and
purportedly have no threshold dose. Hazard evaluation in these
cases is performed by risk estimation. In this approach, mathematical models are used to determine the probability of tumor occurrence in man. However, there are a number of statistical issues that
still need to be resolved before accurate risk estimations can be
made.

Future Research Directions

Risks associated with pesticide use can be more accurately assessed
through a better understanding of the current toxicology data base.
This can be achieved by the development of new strategies designed
to elucidate the biological mechanisms underlying toxicologic
responses. For example, a number of shortcomings of traditional and
present protocols render the prediction of carcinogenic risks to
humans from animal experimentation unsatisfactory. New strategies
should include incorporation of pharmacokinetic and metabolism data
to allow elucidation of genotoxic mechanisms as a basis for better
risk evaluation(4). Use of such data to assist in selection of
dosage levels for chronic studies will ensure that normal metabolic
processes are not overloaded and that biological responses of the
test animals are more representative of the actual toxicity of the
pesticide. This will allow for more precise extrapolation of data
from animals to man.
    A variety of liver culture systems are available for detecting
genotoxic (DNA-damaging) and epigenetic (non DNA-damaging) effects
of carcinogens. Current evidence indicates that carcinogens have
distinctive properties and probably act by different mechanisms(5).
Therefore, different risk assessments could be used for different
types of carcinogens. While conservative one-hit risk models may be
appropriate for some genotoxic carcinogens, other risk models, such
as the Weibull model, may better approximate the risk associated with
epigenetic carcinogens(6,7).
    While the list of animal teratogens grows longer and longer,
few significant advances have been made to increase our understanding of the mechanism of teratogenesis and our ability to extrapolate
these findings to human reproductive hazard assessment. While no
single test species can be said to accurately predict the true human
response to a given test substance, tests in multiple species may
increase the predictive reliability of animal test data. Increased
basic research coupled with better monitoring of human populations
will eventually lead to a better understanding of how animal data
can best be used to preduct human reproductive risks(8).
    Finally, immunotoxicology has been a topic of much interest
today, especially considering the drastic consequences of immunosuppression which can result in the loss of immune surveillance/pro-
tective mechanisms against infectious agents and cancer. However,
no consensus exists on which test procedures are most suitable for
incorporation into routine toxicology assays. One approach which
has been considered is to conduct immunotoxicity testing during
standard subchronic toxicity studies by collecting peripheral blood
samples for immune function tests(9).

Conclusion

New strategies are required for carcinogenicity testing and risk estimation. Current testing requirements involve administration of pesticides to test animals at dosage levels which may be of such magnitude that they substantially alter the animals normal metabolic, physiologic or pharmacologic response. Clearly, doses which result in metabolic overload and are unrealistic, given typical human exposure patterns, should be avoided in chronic bioassays. Results of studies obtained at such doses cannot be reliably extrapolated to the low doses to which man will be exposed. Therefore, risk estimation should not be performed for toxic effects produced in test animals that are physiologically compromised. Also, extrapolation of results from a properly conducted study should be performed using the most appropriate risk model; for example, one which most accurately fits the experimental data and also takes into account the probable mechanisms of action of the test substance.

Improved test methods are needed to better assess potential adverse effects on the reproductive and immune systems. This can only be obtained through increased basic research and critical evaluation of adverse effects which are presently found in human populations. In this manner, results of animal testing can be better extrapolated to man.

Literature Cited

1. "Data Requirements for Pesticide Registration", Environmental Protection Agency, Federal Register, 1984, 49, 42856-905.
2. "Pesticide Assessment Guidelines, Subdivision F Hazard Evaluation: Human and Domestic Animals", Office of Pesticide Programs, Environmental Protection Agency, 1982.
3. Wolf, F. J. J. Environ. Path. Toxicol. 1980, 3, 113-134.
4. Henschler, D. Trends in Pharm. Sci. (FEST Supplement) 1985, 6, 26-8.
5. Williams, G. M. Reg. Toxicol. Pharm. 1985, 5, 132-44.
6. Carlborg, F. W. Fd. Cosmet. Toxicol. 1981, 19, 155-63.
7. Rodricks, J.; Taylor, M. R. Reg. Toxicol. Pharm. 1983, 3, 275-307.
8. Frankos, V. H. Fund. Appl. Toxicol. 1985, 5, 615-25.
9. Noebury, K. C. J. Am. College Toxicol. 1985, 4, 279-90.

RECEIVED September 11, 1986

Chapter 3

Acute Versus Chronic Toxicity and Toxicological Interactions Involving Pesticides

Raymond S. H. Yang

National Toxicology Program, National Institute of Environmental Health Sciences,
P.O. Box 12233, Research Triangle Park, NC 27709

Selected examples of insecticides, herbicides, fungicides, nematocides and fumigants are discussed in
regard to the similarities or differences in their
respective acute, subchronic, and chronic toxicity.
Similarity, in the sense of toxicity to the same target
organ, is more likely an exception rather than the
rule. For the purpose of this symposium, the area of
toxicological interaction involving pesticidal chemicals is of relevance. The classical examples of
synergism and/or potentiation between such binary mixtures as pyrethrin and piperonyl butoxide or malathion
and EPN are well known. A review is given on some of
the more recent examples of modulation of toxicity of
chemicals involving pesticides. Finally, an approach
is suggested to deal with the increasingly complex
problems in toxicology.

Since this chapter is primarily dealing with the comparative aspects
of acute and chronic toxicity of pesticidal chemicals, it is
appropriate to begin with a discussion of the differences between
acute and chronic toxicity. When one reads the readily available
books, monographs, and other documents in toxicology (1-9), it is
quite easy to find a clear-cut definition for acute toxicity, but not
for chronic toxicity. For instance, the Organisation of Economic
Cooperation and Development (OECD), in its Guideline for Testing
Chemicals (5), defined acute toxicity as "the adverse effects
occurring within a short time of administration/exposure of a single
dose of a substance or multiple doses given in 24 hours". An actual
definition was not given in the same document for chronic toxicity.
In the proposed Health Effects Test Standards for Toxic Substances
Control Act Test Rules by the EPA (6), "chronic effects" was defined
as "...disease processes which have a long latency period for development, result from long-term exposure, are long-term illnesses, or
combinations of these factors." But ambiguity still exists; for
instance, how long is "long-term"? The reason for the absence of a
clear-cut definition for chronic toxicity is the complexity of events

and factors involved in the development of chronic toxicity in animals. In Table I, a comparison of acute and chronic toxicity is
given according to a variety of factors.

Table I. Comparison of Acute and Chronic Toxicity

Factors	Acute Toxicity	Chronic Toxicity
Exposure		
Frequency	Single/Repeated/ Continuous	Repeated and prolonged
Duration	Within 24 hrs	At least $1/2$ of the life span; less in humans
Pharmacokinetics	Blood level high; bioavailable for a short time (i.e. hrs, days)	Gradual build up of blood level; bioavaila-bility prolonged (i.e. months, yrs)
Responses	Immediate or in a short time (i.e. within days) Involving few target organs/ systems	Delayed; prolonged (i.e. months, yrs); may be self-propagating Diversified target organs/systems
Experimental design	Many differences involving a number of para-meters; see text	

References (1-9)

It must be noted at the outset that certain grey areas do exist
between acute and chronic toxicity and a sharp distinction may not be
drawn between the two in certain cases. More detailed discussion on
specific examples will be given later.

With respect to the frequency and duration of exposure, as indicated before, acute toxicity is the result of single or repeated or
continuous exposure within a 24-hr period. On the other hand, chronic exposures, which are different from chronic effects (1,9), are
defined as repeated or continuous exposures over a long period of
time, generally more than half of the life span of the animal,
although a shorter period is used for humans (9). Figure 1, although
an oversimplification, illustrates the theoretical pharmacokinetic
differences between acute and chronic exposure. In general, acute
exposure, because of higher dosages, would result in higher blood
levels in a relatively short time span. Chronic exposure, on the
other hand, involves relatively low dosage levels with accumulation
over a long period of time. If one compresses these blood kinetic
profiles to the actual scales of a subchronic or chronic exposure,
the theoretical differences become immediately apparent.

In Figure 1, every rise and fall reflects a dosing interval.
When the time scale becomes "years", the blood level of the chemical
may only be represented graphically as a solid block between the
minimal and maximal concentrations.

With respect to toxicological responses or effects, there may be
quantitative or qualitative differences between acute and chronic

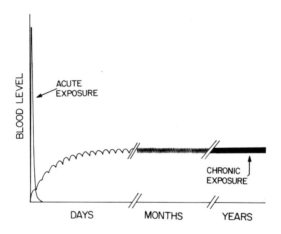

Figure 1. Theoretical pharmacokinetic profiles of a chemical in the blood of a mammal dosed via acute, subchronic, or chronic exposures in relation to different time scales.

toxicity. For many chemicals, the acute toxic effects are different from the toxicities resulting from chronic exposure (4,10). Chan et al. (11) summarized the common signs and the organs, tissues, or systems most likely to be involved in acute toxicity studies. An overwhelmingly large portion of the acute toxic effects was primarily related to the dysfunction of the nervous system. On the other hand, if one considers only one specific area of chronic toxicity, chemical carcinogenesis, the target sites may include hematopoietic, respiratory, digestive, endocrine, urinary, and reproductive systems comprising many different organs.

Two other points that deserve special mention are the self-propagating effects of certain types of chronic toxicity (i.e., neoplasm, advanced stages of cirrhosis) and the long, latent period for certain instances of chronic toxicity (i.e., delayed neurotoxicity, neoplasms). This latter category of chronic toxicity may be induced in some instances after only one single exposure. In experimental toxicology, the designs of acute and chronic toxicity studies vary greatly in regard to animal numbers, dosages, environmental conditions, modes and routes of exposure, duration of the studies, toxicological endpoints, and statistical analyses. These differences are beyond the scope of this presentation.

Acute and Chronic Toxicity of Pesticides

There are literally thousands of chemicals and/or formulations in the major categories (i.e., insecticides, herbicides, fungicides, rodenticides, fumigants, nematocides) of pesticides. Therefore, no attempt was made to provide a review of representatives of all the major classes of pesticides. In the section which follows, selected pesticides from three chemical classes, the organophosphates, the halogenated hydrocarbons, and the pyrethroids, will be discussed in regard to the differences and similarities between acute and chronic toxicity. The criteria for selection of the examples are mainly related to the availability of current information in the literature.

Organophosphates. The acute toxicity of organophosphate pesticides is basically derived from the anticholinesterase property of these chemicals. This property, which results in accumulation of acetylcholine at synapses and myoneural junctions, is responsible for both the insecticidal activity and mammalian toxicity. Early symptoms of organophosphate poisoning in humans include, among others, miosis (pinpoint pupils) and blurred vision, and a response known as the SLUD (salivation, lacrimation, urination, and diarrhea) syndrome; all of these are the result of muscarinic effects (12-15). Clinical manifestations of more severe poisoning involve predominantly nicotinic and central effects which include convulsions, paralysis, depressed respiration and cardiovascular functions, and coma (12-15). Death is usually due to respiratory failure, accompanied by cardiovascular failure (13).

Chronic toxicity of organophosphates may be discussed under four different areas: carcinogenicity, delayed neurotoxicity, experimental myopathy, and, in humans, psychiatric disorders. In 1983, the IARC (International Agency for Research on Cancer) evaluated the carcinogenic potential of, among other pesticides, five organophosphate insecticides/acaricides (malathion, methyl parathion, parathion,

tetrachlorvinphos, and trichlorfon) (16). One of these, tetrachlorvinphos, was considered to have provided limited evidence of carcinogenicity to experimental animals. This evaluation was based on
the treatment-related increases of hepatocellular adenoma and car-
cinoma in B6C3F$_1$ mice, and C-cell adenoma of the thyroid and cortical
adenoma of the adrenal in Osborne-Mendel rats (16). A recent oncogenicity study of tetrachlorvinphos in B6C3F$_1$ mice confirmed the carcinogenic potential of this chemical to the male mice at a very high
dose level (i.e., 16,000 ppm, the highest level used in an earlier
NCI study) but the investigators questioned the utility of results
from a dose level which exceeded the maximum tolerated dose level
(17). The four other organophosphates evaluated by the IARC were
reported as either having inadequate evidence to evaluate its carcinogenicity to experimental animals (parathion, trichlorfon) or that
the available data did not provide evidence for carcinogenicity in
experimental animals (malathion, methyl parathion). Two recent
publications (18,19) which appeared in the same issue of
Environmental Research raised a controversy over the carcinogenicity
of malathion and its oxygen analog, malaoxon. Reuber (18), a pathol-
ogist claimed to have made the histopathological examination of the
National Cancer Institute (NCI) studies, felt strongly that both
malathion and malaoxon are carcinogenic in Osborne-Mendel and Fischer
344 rats. However, the NCI Technical Reports (20-22), which did not
list Reuber as a contributor, concluded that neither chemical was
shown to be carcinogenic. The other publication, by Huff et al.
(19), summarized the results of histopathology reexamination by a
team of pathologists at and/or for the National Toxicology Program
(NTP) following the renewed public health interests and concern about
the increasing use of malathion in agriculture and especially its use
to control Mediterranean fruit fly in California and Florida during
the 1980s. The NTP team confirmed the original NCI interpretative
conclusion that malathion was not carcinogenic. For the malaoxon
study, the only difference between the original and subsequent
interpretations was for C-cell neoplasms of the thyroid gland, in
that the NTP concluded that there was equivocal evidence of carcinogenicity for male and female F344 rats (19).
      The delayed neurotoxicity, experimental myopathy, and
psychiatric disorders (human) represent some of the grey areas between acute and chronic toxicity. Since these effects may be induced
with one or few exposures with a short latency period, they should be
considered as acute effects. However, they are grouped under chronic
toxicity in this presentation for three reasons: (a) there is reason
to believe that at least one and possibly more of these toxic responses may be induced chronically; (b) the exposure periods to humans in
clinical cases were often uncertain; (c) these responses are all considered as chronic effects in the medical literature (e.g. 14,15).
      Delayed neurotoxicity results from degeneration of the axons
followed by demyelination (14,15,23). Clinical manifestation includes sensory disturbances, ataxia, weakness, muscle twitching and,
in severe cases, complete flaccid paralysis (15). A fair number of
organophosphate compounds are capable of inducing delayed neurotoxicity. Of the 250 organophosphates (not all pesticides) tested for
delayed neurotoxicity in chickens, 47% (117 chemicals) showed positive responses (23). Notable examples of pesticides which possess
this neurotoxicity are leptophos, EPN, merphos, dichlorvos, and

trichlorfon (23). In general, delayed neurotoxicity may be induced
by these chemicals after a single dose with a latent period of about
6 to 14 days (23). However, in HSD,ICR mice, two single oral doses
of 1000 mg/kg of TOCP at a 21-day interval failed to induce neuropathy whereas daily dosing of 225 mg/kg TOCP for 9 months caused,
among other toxic signs, muscle wasting, weakness and ataxia which
progressed to severe hindlimb paralysis at the termination of the
study (24). The development of neuropathy in this case was very
slow; severe ataxia was not evident until after about 8 months of
dosing (24). This may very well be an example where a similarity
exists between acute and chronic toxicity.

Experimental myopathy following acute or repeated administration
of certain organophosphates in laboratory animals is characterized
initially by focal necrosis and subsequently by a generalized breakdown of muscle fiber architecture (14,15,25). The diaphragm appears
to be the most severely affected (25). Myopathic alterations in
humans following organophosphate poisoning have also been reported
(14,26,27). Related toxic responses include muscle tenderness,
changes in surface electromyography (EMG) and elevated muscle enzymes
such as CPK (creatine phosphokinase)(14). Psychiatric disorders
including acute psychosis or severe depressions were reported in
greenhouse workers, farm workers and scientists working with organophosphate pesticides (14).

Halogenated Hydrocarbons. Three representative chemicals,
2,4-dichlorophenoxyacetic acid (2,4-D), hexachlorobenzene (HCB) and
1,3-dichloropropene (DCP), will be discussed in this section.

2,4-D, a well-known herbicide, is of low to moderate acute toxicity; depending on the salt forms used, the LD_{50}'s are in the order
of several hundred to about 2000 mg/kg body weight (28). Animals
given a lethal dose of 2,4-D appear to die from ventricular fibrilla-
tion. At sublethal doses, toxic responses to 2,4-D are indicative of
neuromuscular involvement, including stiffness of the extremeties,
ataxia, paralysis and eventually coma (29-31). The central nervous
system appeared to be a target organ in the acute toxicity of 2,4-D
(31). A number of chronic toxicity studies were reviewed recently by
Collins (31). Most of the experiments revealed no 2,4-D treatment-
related chronic toxicity. In 1982, the IARC considered that there
was inadequate evidence for carcinogenicity of 2,4-D to humans and
animals (32) and that status is still true in 1986 (31). However,
there is a report in which neurobehavioral signs, as well as changes
in clinical chemistry parameters, were observed in pigs fed the
triethanolamine salt of 2,4-D at 500 mg/kg of diet for up to one year
(33). Kidney lesions (epithelial regeneration) of minimal severity
were seen in male Fischer 344 rats on a 13-week subchronic study
sponsored by The Industry Task Force on 2,4-D Research Data (31).

HCB has been used as a fungicide to control wheat bunt and smut
fungi on other grains (34). However, the major source of environmental concern for HCB is derived from its being a byproduct or waste
material in the production of many chemicals (34,35). HCB has a low
order of acute toxicity; its oral LD_{50} ranges from 1,000 to 10,000
mg/kg in several animal species (34,35). Toxic signs as a result of
acute exposure to HCB are related to neurotoxic manifestations such
as trembling, ataxia and paralysis (36). Death is due to neurotoxic
effects (34,37).

Chronic toxicity of HCB involves three different areas: carcinogenicity, porphyria, and neurotoxicity. A number of carcinogenicity studies were conducted with Syrian golden hamsters,
Swiss mice, and Wistar and Sprague-Dawley rats, and they were
reviewed recently (35,38,39). HCB is carcinogenic in all three species, and the major target organs include liver, kidney, thyroid,
parathyroid, adrenal, and the lymphohematopoietic system (35,38,39).
Experimental porphyria may be induced by a single dose of HCB
(40) as well as by repeated or chronic exposure to HCB (34,35,40),
another example of similarity between acute and chronic toxicity. An
epidemic of about 4000 cases of human porphyria (porphyria cutanea
tarda or porphyria turcica) occurred in Turkey between 1955 and 1959
as a result of consumption of grain that had been treated with HCB
(34,35,41,42). Clinical manifestations included generalized hyper-
pigmentation and hypertrichosis, scarring on the cheeks and hands,
and tight sclerodermoid changes of the nose with perioral scarring;
and in children, painless arthritic changes with osteoporosis of car-
pal, metacarpal and phalangeal bones and atrophy or failure to develop in the terminal phalanges. In addition, neurologic symptoms
including weakness, paresthesias, myotonia, and cogwheeling were
observed (34,35,41,42). Neurotoxic signs such as ataxia and paralysis have also been observed in experimental animals treated with HCB
subchronically or chronically (35). It is noteworthy that other
toxic effects of HCB, including immunosuppression, body and organ
weight changes, alterations of clinical pathology parameters, were
also reported (34,35).
DCP is the main ingredient of a number of commercial fumigant
formulations. Information on the toxicology of DCP is generally
derived from the results of the commercial preparation Telone II
(approximately 92% DCP) and/or D-D (approximately 52% DCP) (43).
DCP, Telone II and D-D are moderately toxic to mammals in acute exposures. The primary target organs in rats following acute oral dosing
of Telone are liver, kidneys and possibly lung (44). The acute toxic
responses in rats and mice following peroral treatment of D-D
include hyperexcitability, followed by tremors, incoordination,
depression, and dyspnea. The pathological changes in animals that
died from D-D exposure included distension of the stomach by fluid
and gas, erosion of the gastrointestinal mucosa, occasional hemor-
rhages in the lungs, and fatty degeneration of the liver (45).
Chronic toxicity of Telone II following gavage dosing (3 times/week)
for two years in Fischer 344 rats and/or B6C3F$_1$ mice included
neoplasms (forestomach, liver, lung, urinary bladder), epithelial
hyperplasia (forestomach, urinary bladder) and hydronephrosis
(43,46,47).

Pyrethroids. Pyrethroids, such as natural pyrethrins and synthetic
analogs, allethrin, permethrin, and others, are well known for their
neurotoxicity (48-59). However, as a major class of insecticide,
they have a remarkable safety margin for mammals, principally because
of the rapid metabolic degradation of pyrethroids in mammalian species (48-50). The acute toxicity of pyrethroids involves two
distinct syndromes in rats and mice (49-51). The first one, T
syndrome or tremor (Type I), is characterized by a rapid onset of
tremor, initially in the limbs and gradually extending over the whole
body. Death is associated with clonic seizures. The second

syndrome, CS-syndrome or choreoathetosis with salivation (Type II), is characterized by profuse salivation followed by a gradual development of a coarse whole-body tremor and a splayed gait of the hind legs. Clonic and tonic seizures in both species usually result in death.

Despite their popularity as a major class of insecticides, information on chronic toxicity of pyrethroids is limited. Of the several studies mentioned or reported (49,50,52-57), there appeared to be no evidence for carcinogenicity nor were there any significant morphological or ultrastructural alterations in the nervous system. Neurotoxic signs were observed in some of the studies, but they either disappeared after a short period of time or were not accompanied by any significant neuropathy (53,55). Hepatic changes as a result of repeated or chronic exposure to pyrethroids were reported quite consistently (49,52-55). These changes included hypertrophy, bile duct proliferation and multifocal microgranulomata (52-55). Multifocal microgranulomata were also observed in lymph nodes and spleen in mice exposed to fenvalerate chronically (54).

Toxicological Interactions Involving Pesticides

The classical examples of synergism and/or potentiation between such binary mixtures as pyrethrin and piperonyl butoxide or malathion and EPN are well known (60,61). The effects of solvents and impurities on the insecticidal activity of commercial preparations has also been studied (62,63). In a number of recent articles (64-68), toxicological interactions of pesticides and other chemicals were discussed with respect to specific target organs, to dietary and nutritional factors and in light of the possibility of designing safer chemicals. Since, without an exception, all pesticidal chemical formulations are chemical mixtures, the following two examples are presented in the hope of bringing additional attention to this very important area of toxicology.

Mehendale and associates have conducted a series of studies on the effects of pretreatment of Kepone (chlordecone) on the acute toxicity of carbon tetrachloride (69-79). After the discovery of the dramatic potentiation of carbon tetrachloride hepatotoxicity by pretreatment of Kepone (69), Klingensmith and Mehendale (73) and Mehendale (74) demonstrated that pretreatment of young male Sprague-Dawley rats with a very low level (10 ppm) of Kepone in the diet for 15 days enhanced acute toxicity of carbon tetrachloride 67-fold (Table II).

This is probably the first report where a chemical at an environmentally realistic level (i.e. 10 ppm) caused a dramatic potentiation/synergism in the toxicity of another chemical. This potentiative or synergistic effect of Kepone was apparently rather specific in that close structural analogs such as mirex and photomirex do not share this property (74). The underlying mechanism for this phenomenon is being pursued actively in Dr. Mehendale's laboratory, it probably involves the excessive accumulation of intracellular calcium ion and the disruption of hepatocellular repair-regeneration processes (77-79).

Toxicological interactions may occur in chronic toxicity and carcinogenicity studies (80,81). Wong et al. (80) examined the influence of disulfiram on the chronic toxicity of EDB (ethylene

Table II. Enhancement of Acute Toxicity of Carbon Tetrachloride by
           Low Level Dietary Pretreatment of Kepone

Dietary Pretreatment	48 hr LD_{50} (ml/kg)	Increase in Mortality
Carbon tetrachloride		
Control diet	2.8	--
Kepone (10 ppm) diet	0.042	67-fold

Condensed from Klingensmith and Mehendale (73) and Mehendale (74)

dibromide) in a long-term inhalation study. This work came about
because the National Institute of Occupational Safety and Health had
been interested in minimizing hazards related to workers' exposure to
EDB, particularly to those who were in alcohol control programs under
disulfiram (antabuse) treatment. The rationale for this concern was
derived from the enzyme inhibitory properties of disulfiram toward
acetaldehyde dehydrogenase. Since this enzyme plays a key role in
the biotransformation of EDB, it was thought that its inhibition
might modify the toxicity of EDB, including its carcinogenicity.

Table III. Experimental Design of the EDB/Disulfiram Interaction
            Study

Animal:	Sprague-Dawley rats
Test groups:	Control
	EDB (20 ppm, inhalation)
	Disulfiram (0.05% in diet)
	EDB + Disulfiram
Group size:	48 rats/group/sex
Duration of exposure:	18 months
Exposure frequency:	7 hr/day, 5 days/week (inhalation)
	Diet given ad libitum except
	chamber exposure period
Endpoints	Body weights
	Food consumption
	Mortality
	Hematology
	Gross and Histopathology

Condensed from Wong et al. (80)

    The experimental design of the study by Wong et al. (80) is sum-
marized in Table III. Dramatic enhancement of mortality was observed
as early as 9 months in the EDB/Disulfiram combination group and the
results of the cumulative mortality of the last 9 months of the study
are summarized in Table IV.

Table IV. Cumulative Mortality in Rats of the EDB/Disulfiram
Interaction Study

		Male				Female		
	9	12	15	18	9	12	15	18
Control	0	0	1	5	0	2	3	6
Disulfiram	1	2	4	6	1	2	2	3
EDB	1	5	30	43	1	4	19	37
EDB + Disulfiram	8	23	48	48	12	40	48	48

Condensed from Wong et al. (80)

Toxicological interactions were seen also in the marked increases of
tumor incidences (except mammary tumors) as well as in the shortening
of the latent period. Table V is a summary of the major histopathological findings in the EDB alone group and the EDB/disulfiram combination group. Control and disulfiram alone groups were not
included to conserve space. Other than the significant increase of
mammary tumors in the disulfiram alone female rats, no other tumor
incidences were different from the background in these two control
groups.

Table V. Major Histopathological Findings in Rats Exposed to EDB or
EDB/Disulfiram (EDB+DS) in the EDB/Disulfiram
Interaction Study

	EDB		EDB+DS	
	Male	Female	Male	Female
No. of Animals Examined	46	48	48	45
Liver				
hepatocellular tumors	2	3	36*	32*
Mesentary or omentum				
hemangiosarcoma	0	0	11*	8*
Kidney				
adenoma and adeno-				
carcinoma	3	1	17*	7*
Thyroid				
follicular epithelial				
adenoma	3	1	18*	18*
Mammary				
all tumors	---	25	---	13*
Lung				
all tumors	3	0	9*	2
No. of Rats with Tumor	25	29	45*	45*
No. of Rats with				
Multiple Tumors	10	8	37*	32*

* $P < 0.05$

Condensed from Wong et al. (80)

A Possible Approach For Dealing With Toxicological Interactions

Having reviewed these two examples of toxicological interactions
involving pesticides, let us put things into perspective. Are we in
imminent danger because of potential toxicological interactions due
to environmental pollution or occupational exposure to pesticides
and/or other chemicals? The answer is probably "no". Dr. John
Doull stated some reasons why the answer is no in the Proceedings of
the 5th International Congress of Pesticide Chemistry "...As a group,
the pesticides have been subjected to a more thorough toxicologic
investigation in animals than any other class of chemicals, including
drugs..." and "...Considering the extreme toxicity of some of the
pesticides and the severity of their adverse effects, their overall
safety record is remarkably good..." (82). Should we then ignore the
issue of toxicological interactions? The answer is obviously a "no"
because, as long as we use these chemicals to enhance the quality of
our life, we can not possibly afford to have a detrimental surprise
in the human population. What are we to do then? The long-range
answer, in my view, lies in the application of some of the recent
advances in pharmacokinetics and computer technology.

As shown in Figure 2, the concepts of "physiologically-based
pharmacokinetics" and "animal scale-up" were initiated in the late
1960s and early 1970s by Bischoff et al. (83-85), Dedrick et al.
(86), and Dedrick (87,88). Physiologically-based pharmacokinetics
differ from the classical pharmacokinetics in that: (a) the utiliza-
tion of a large body of physiological and physicochemical data which
are not chemical specific; (b) interspecies extrapolation may be
attempted with more confidence; (c) the pharmacokinetic behavior of
certain chemicals may be predicted a priori or from very limited
data; and (d) compartments correspond to anatomical entities such
that organ or tissue specific biochemical interactions can be incorporated (88).

If one draws an analogy of the "scale-up" from a mouse to a
human to the scale-up of a chemical plant or a chemical engineering
process, one finds that both situations are governed by a great
number of physical and chemical processes (87). In mammals, the
physical processes (i.e., mass balances, thermodynamics, transport,
and flow) often vary in a predictable way. However, chemical processes such as metabolic reactions may vary greatly and unpredictably
among species. The physical and chemical processes interact such
that the pharmacokinetics of any given chemical between one species
and another may be predictable depending on the amount of background
information available (87).

In the past, the application of physiologically-based pharmacokinetics was limited by the complexity of the mathematics
involved because of the large number of parameters in the models. In
recent years, the advances in computer software have overcome this
limitation. Thus, earlier this year, Clewell and Andersen (89)
reported that by using the Advanced Continuous Simulation Language
(ACSL), physiologically-based pharmacokinetic modelling may be
carried out on personal computers with reasonably short turn-around
times (i.e., execution time, 0.6-8 minutes) and in a user-friendly
manner.

Presently, the application of these techniques in toxicology is
being actively pursued for the extrapolation between routes, between

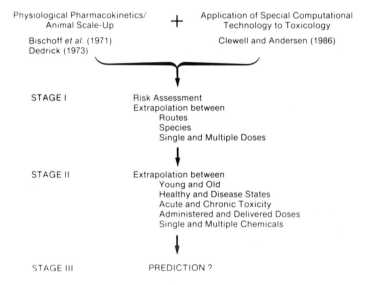

Figure 2. A suggested approach utilizing physiologically based pharmacokinetics and computer technology for the extrapolation and prediction of various situations in toxicology.

species, between single and multiple doses, and in its application to
the risk assessment process (90-95). All these efforts are combined
in Figure 2 as a Stage I effort. It is conceivable that in the fore-
seeable future, with the availability of more research data, further
extrapolation may be made between the young and the old, the healthy
and the diseased, acute and chronic toxicity, the administered
(exposure) and delivered (effective) doses, and between single and
multiple chemical exposures (Stage II effort, Figure 2). The ultimate hope is, of course, to "predict" the possible outcome of toxicity for one or more chemicals (including toxicological interactions)
in one or more species with little or no need of doing the actual
tedious animal experimentation such as the chronic toxicity studies.

Acknowledgments

I thank Drs. K. M. Abdo, J. E. Huff, E. E. McConnell, J. K. Selkirk
of NIEHS, Dr. M. E. Andersen of Wright-Patterson Air Force Base, Dr.
H. M. Mehendale of the University of Mississippi Medical Center, and
Dr. R. J. Kuhr of North Carolina State University for their critical
review of the manuscript.

Literature Cited

1. Doull, J.; Klaassen, C. D.; Amdur, M. O. "Casarett and Doull's
    Toxicology. The Basic Science of Poisons"; 2nd Edition;
    Macmillan Publishing Co., Inc.: New York, 1980; 778 pp.
2. Hayes, A. W. "Principles and Methods of Toxicology"; Student
    Edition; Raven Press: New York, 1984; 750 pp.
3. Loomis, T. A. "Essentials of Toxicology"; 3rd Edition; Lea &
    Febiger: Philadelphia, 1978; 245 pp.
4. Klaassen, C. D. In "Goodman and Gilman's The Pharmacological
    Basis of Therapeutics"; 6th Edition; Gilman, A. G.; Goodman, L.
    S.; Gilman, A.; Mayer, S. E.; Melmon, K. L., Eds.; Macmillan
    Publishing Co., Inc.: New York, 1980; Chap. 68.
5. "OECD Guidelines for Testing of Chemicals," Organisation for
    Economic Co-operation and Development, 1981.
6. "Proposed Health Effects Test Standards for Toxic Substances
    Control Act Test Rules," U. S. Environmental Protection Agency,
    Fed. Reg. 1979, 44, 27334-375.
7. "Proposed Guidelines for Registering Pesticides in the United
    States; Hazard Evaluation: Human and Domestic Animals," U. S.
    Environmental Protection Agency, Fed. Reg. 1978, 43, 37336-403.
8. "Principles and Procedures for evaluating the Toxicity of
    Household Substances," National Academy of Sciences, 1977.
9. "Drinking Water and Health," Vol. 1, National Academy of
    Sciences, 1977.
10. Klaassen, C. D.; Doull, J. In "Casarett and Doull's Toxicology.
    The Basic Science of Poisons"; 2nd Edition; Doull, J.; Klaassen,
    C. D.; Amdur, M. O., Eds.; Macmillan Publishing Co., Inc.: New
    York, 1980; Chap. 2.
11. Chan, P. K.; O'Hara, G. P.; Hayes, A. W. In "Principles and
    Methods of Toxicology"; Student Edition; Hayes, A. W., Ed.;
    Raven Press: New York, 1984; Chap. 1.
12. Haddad, L. M. In "Clinical Management of Poisoning and Drug
    Overdose"; Haddad, L. M.; Winchester, J. F., Eds.; W. B.
    Saunders: Philadelpnia, 1983; Chap. 67-68.

13. Gossel, I. A.; Bricker, J. D. "Principles of Clinical Toxicology"; Raven Press: New York, 1984; Chap. 9.
14. Stopford, W. In "Industrial Toxicology. Safety and Health Applications in the Workplace"; Williams, P. L.; Burson, J. L., Eds.; Van Nostrand Reinhold Co.: New York, 1985; Chap. 11.
15. Taylor, P. In "Goodman and Gilman's The Pharmacological Basis of Therapeutics"; 6th Edition; Gilman, A. G.; Goodman, L. S.; Gilman, A.; Mayer, S. E.; Melmon, K. L., Eds.; Macmillan Publishing Co., Inc.: New York, 1980; Chap. 6.
16. "IARC Monographs on the Evaluation of the Carcinogenic Risk of Chemicals to Humans. Miscellaneous Pesticides," International Agency for Research on Cancer, 1983, Vol. 30.
17. Parker, C. M.; Van Gelder, G. A.; Chai, E. Y.; Gellatly, J. B. M.; Serota, D. G.; Voelker, R. W.; Vesselinovitch, S. D. Fundam. Appl. Pharmacol. 1985, 5, 840-54.
18. Reuber, M. D. Environ. Res. 1985, 37, 119-153.
19. Huff, J. E.; Bates, R.; Eustis, S. L.; Haseman, J. K.; McConnell, E. E. Environ. Res. 1985, 37, 154-173.
20. "Bioassay of Malathion for Possible Carcinogenicity, CAS No. 121-75-5," Technical Report Series, No. 24, National Cancer Institute, 1978.
21. "Bioassay of Malaoxon for Possible Carcinogenicity, CAS No. 1634-78-2," Technical Report Series, No. 135, National Cancer Institute, 1979.
22. "Bioassay of Malathion for Possible Carcinogenicity, CAS No. 121-75-5," Technical Report Series, No. 192, National Cancer Institute, 1979.
23. Abou-Donia, M. B. Ann. Rev. Pharmacol. Toxicol. 1981, 21, 511-48.
24. Lapadula, D. M.; Patton, S. E.; Campbell, G. A.; Abou-Donia, M. B. Toxicol. Appl. Pharmacol. 1985, 79, 83-90.
25. Laskowski, M. B.; Dettbarn, W. D. Ann. Rev. Pharmacol. Toxicol. 1977, 17, 387-409.
26. De Reuck, J.; Willems, J. J. Neurol. 1975, 208, 309-14.
27. Wecker, L.; Mrak, R. E.; Dettbarn, W. D. J. Environ. Pathol. Toxicol. Oncol. 1985, 6, 171-6.
28. "IARC Monographs on the Evaluation of the Carcinogenic Risk of Chemicals to Humans. Some Fumigants, the Herbicides 2,4-D and 2,4,5-T, Chlorinated Dibenzodioxins and Miscellaneous Industrial Chemicals," International Agency for Research on Cancer; Vol. 15; 1977.
29. Gehring, P. J.; Betso, J. E. Ecol. Bull. 1978, 27, 122-33.
30. Hill, E. V.; Carlisle, H. J. Ind. Hyg. Toxicol. 1947, 29,85-95.
31. Collins, J. J. Rev. Environ. Contam. Toxicol. Manuscript submitted.
32. "IARC Monographs on the Evaluation of the Carcinogenic Risk of Chemicals to Humans," International Agency for Research on Cancer, 1982, p. 101, Supplement 4.
33. Bjorkland, N. E.; Erne, K. Acta Vet. Scand. 1966, 7, 364-90.
34. "IARC Monographs on the Evaluation of the Carcinogenic Risk of Chemicals to Humans. Some Halogenated Hydrocarbons," International Agency for Research on Cancer, 1983, Vol. 20, pp. 155-78.
35. "Health Assessment Document for Chlorinated Benzenes," U.S. Environmental Protection Agency, 1985.

36. Gehring, P. J.; MacDougall, D. "Review of the Toxicity of
     Hexachlorobenzene and Hexachlorobutadiene;" Dow Chemical,
     U.S.A.; 1971.
37. Booth, N. H.; McDowell, J. R. J. Amer. Vet. Med. Assoc. 1975,
     166, 591-5.
38. Cabral, J. R. P.; Shubik, P. International Symposium on
     Hexachlorobenzene, International Agency for Research on Cancer,
     Lyon, France, June 24-28, 1985. (Abstract).
39. Erturk, E.; Lambrecht, R. W.; Peters, H. A.; Morris, C. R.;
     Bryan, G. T. International Symposium on Hexachlorobenzene,
     International Agency for Research on Cancer, Lyon, France, June
     24-28, 1985. (Abstract).
40. Sweeney, G. D.; Janigan, D.; Mayman, D.; Lai, H. S. A. J. Lab.
     Clin. Med. 1971, 17, 68-72.
41. Goemen, A. International Symposium on Hexachlorobenzene,
     International Agency for Research on Cancer, Lyon, France, June
     24-28, 1985. (Abstract).
42. Peters, H. A.; Cripps, D. J.; Goemen, A.; Erturk, E.; Bryan, G.
     T.; Morris, C. R. International Symposium on Hexachlorobenzene,
     International Agency for Research on Cancer, Lyon, France, June
     24-28, 1985. (Abstract).
43. Yang, R. S. H. Residue Rev. 1986, 97, 19-35.
44. Torkelson, T.; Oyen, F. Amer. Ind. Hyg. Assoc. J. 1977, 38,
     217-23.
45. Hines, C. H.; Anderson, H. H.; Moon, H. D.; Kodama, J. K.;
     Morse, M.; Jacobsen, N. W. Arch. Ind. Hyg. Occup. Med. 1953, 7,
     118.
46. "Toxicology and Carcinogenesis Studies of Telone II in F344/N
     Rats and B6C3F$_1$ Mice (Gavage Studies)," National Toxicology
     Program, Tech. Rep. No. 269, 1985.
47. Yang, R. S. H.; Huff, J. E.; Boorman, G. A.; Haseman, J. K.;
     Kornreich, M. J. Toxicol. Environ. Health 1986, 18, 377-92.
48. Casida, J. E. Environ. Health Perspect. 1980, 34, 189-202.
49. Casida, J. E.; Gammon, D. W.; Glickman, A. H.; Lawrence, L. J.
     Ann. Rev. Pharmacol. Toxicol. 1983, 23, 413-38.
50. Aldridge, W. N. Proc. 5th Int. Congr. Pesticide Chem., 1982, 3,
     485-90.
51. Gray, A. J. NeuroToxicology 1985, 6, 127-38.
52. Miyamoto, J. Environ. Health Perspect. 1976, 14, 15-28.
53. Litchfield, M. H. Proc. 5th Int. Congr. Pesticide Chem., 1982,
     2, 207-11.
54. Parker, C. M.; McCullough, C. B.; Gellatly, J. B. M.; Johnston,
     C. D. Fundam. Appl. Toxicol. 1983, 3, 114-20.
55. Parker, C. M.; Piccirillo, V. J.; Kurtz, S. L.; Garner, F. M.;
     Gardiner, T. H.; Van Gelder, G. A. Fundam. Appl. Toxicol. 1984,
     4, 577-586.
56. Dyck, P. J.; Shimono, M.; Schoening, G. P.; Lais, A. C.; Oviatt,
     K. F.; Sparks, M. F. J. Environ. Pathol. Toxicol. Oncol. 1984,
     5, 109-117.
57. Parker, C. M.; Patterson, D. R.; Van Gelder, G. A.; Gordon, E.
     B.; Valerio, M. G.; Hall, W. C. J. Toxicol. Environ. Health
     1984, 13, 83-97.
58. Hallenbeck, W. H.; Cunningham-Burns, K. M. In "Pesticides and
     Human Health"; Springer-Verlag: New York, 1985; 166 pp.

59. Parker, C. M.; Albert, J. R.; Vaan Gelder, G. A.; Patterson, D. R.; Taylor, J. L. Fundam. Appl. Toxicol. 1985, 5, 278-286.
60. Murphy, S. D. In "Casarett and Doull's Toxicology. The Basic Science of Poisons"; 2nd Edition; Doull, J.; Klaassen, C. D.; Amdur, M. O., Eds.; Macmillan Publishing Co., Inc.: New York, 1980; Chap. 16.
61. Matsumura, F. "Toxicology of Insecticides"; Second Edition; Plenum Press: New York, 1985; 598 pp.
62. Brattsten, L. B.; Wilkinson, C. F. Science 1977, 196, 1211-3.
63. Ryan, D. L.; Fukuto, T. R. Pestic. Biochem. Physiol. 1985, 23, 413-24.
64. Wilkinson, C. F.; Murphy, M. Drug Metab. Rev. 1984, 15, 897-917.
65. Hook, J. B.; Serbia, V. C. Proc. 5th Int. Congr. Pesticide Chem., 1982, 3, 515-20.
66. Charbonneau, S. M.; Munro, I. C. Proc. 5th Int. Congr. Pesticide Chem., 1982, 3, 521-5.
67. Kaloyanova, F.; Tasheva, M. Proc. 5th Int. Congr. Pesticide Chem., 1982, 3, 527-9.
68. Kaloyanova, F. In "Health Effects of Combined Exposures to Chemicals in Work and Community Environments"; Proceedings of a Course; European Cooperation on Environmental Health Aspects of the Control of Chemicals-Interim Document 11; World Health Organization: Copenhagen, 1983, pp. 165-95.
69. Curtis, L. R.; William, W. L.; Mehendale, H. M. Toxicol. Appl. Pharmacol. 1979, 51, 283-93.
70. Curtis, L. R.; Mehendale, H. M. Drug Metab. Dispos. 1980, 8, 23-7.
71. Agarwal, A. K.; Mehendale, H. M. Fundam. Appl. Toxicol. 1982, 2, 161-7.
72. Agarwal, A. K.; Mehendale, H. M. Toxicology 1983, 26, 231-42.
73. Klingensmith, J. S.; Mehendale, H. M. Toxicol. Lett. 1982, 11, 149-154.
74. Mehendale, H. M. Fundam. Appl. Toxicol. 1984, 4, 295-308.
75. Lockard, V. G.; Mehendale, H. M.; O'Neal, R. M. Exp. Mol. Pathol. 1983, 39, 230-45.
76. Lockard, V. G.; Mehendale, H. M.; O'Neal, R. M. Exp. Mol. Pathol. 1983, 39, 246-56.
77. Agarwal, A. K.; Mehendale, H. M. Toxicol. Appl. Pharmacol. 1986, In press.
78. Bell, A. N.; Lockard, V. G.; Young, R. A.; Mehendale, H. M. Fourth Int. Congr. Toxicol. Tokyo, Japan; July 21-25, 1986. (Abstract).
79. Bell, A. N.; Mehendale, H. M. FASEB, April, 1986. (Abstract).
80. Wong, L. C. K.; Winsteon, J. M.; Hong, C. B.; Plotnick, H. Toxicol. Appl. Pharmacol. 1982, 63, 155-165.
81. Huff, J. E. Environ. Health Perspect. 1983, 47, 359-63.
82. Doull, J. Proc. 5th Int. Congr. Pesticide Chem., 1982, 3, 433-6.
83. Bischoff, K. B.; Brown, R. G. Chem. Eng. Prog. Symp. Ser. No. 66 1966, 62, 32-45.
84. Bischoff, K. B.; Dedrick, R. L.; Zaharko, D. S. J. Pharm. Sci. 1970, 59, 149-54.
85. Bischoff, K. B.; Dedrick, R. L.; Zaharko, D. S.; Longstreth, J. A. J. Pharm. Sci. 1971, 60, 1128-33.
86. Dedrick, R. L.; Bischoff, K. B.; Zaharko, D. S. Cancer Chemotherapy Rep. Part I 1970, 54, 95-101.

87. Dedrick, R. L. In "Pharmacology and Pharmacokinetics"; Teorell, T; Dedrick, R. L.; Condliffe, P. G., Eds.; Plenum Publishing Corp.: New York, 1973; pp. 117-45.
88. Dedrick, R. L. J. Dynamic Syst. Measurement Cont. September, 1973, pp. 255-258.
89. Clewell, H. J.; Andersen, M. E. Proc. Soc. Computer Simulation Winter Multi-Conf., San Diego, CA, Jan. 23-7, 1986.
90. Andersen, M. E.; Clewell, H. J. III; Gargas, M. L.; Smith, F. A.; Reitz, R. H. Toxicol. Appl. Pharmacol. Submitted.
91. "Drinking Water and Health," Vol. 6, National Academy of Sciences, 1986, Chap. 6.
92. Clewell, H. J., III; Andersen, M. E. Toxicol. Ind. Health 1985, 1, 111-31.
93. Hoel, D. G.; Kaplan, N. L.; Anderson, M. W. Science 1983, 219, 1032-7.
94. Hoel, D. G. In "Toxicological Risk Assessment"; Vol. 1. Clayson, D. B.; Krewski, D.; Munro, I., Eds.; CRC Press, Inc.: Boca Raton, FL, 1985; Chap. 10.
95. Lutz, R. J.; Dedrick, R. L. In "New Approaches in Toxicity Testing and Their Application in Human Risk Assessment"; Li, A. P., Ed.; Raven Press: New York, 1985; pp. 129-49.

RECEIVED October 6, 1986

Chapter 4

New Approaches for the Use of Short-Term Genotoxicity Tests To Evaluate Mutagenic and Carcinogenic Potential

Frederick J. de Serres

Office of the Director, National Institute of Environmental Health Sciences, P.O. Box 12233, Research Triangle Park, NC 27709

Two international collaborative studies sponsored by the
International Program on Chemical Safety to evaluate
short-term tests for genotoxicity have been completed.
The first IPCS study was designed to evaluate in vitro
eukaryotic assay systems for use as a complement to the
Salmonella reverse-mutation assay system. The second
IPCS study was designed to evaluate in vivo assay systems
and their ability to discriminate beween carcinogens and
structurally-related noncarcinogens. The results of both
of these international trials have been used to develop a
unified testing strategy for the evaluation of new test
chemicals to evaluate their mutagenic and carcinogenic
potential.

It was about ten years ago that short-term tests with Salmonella were
first recommended for use in screening environmental chemicals for
mutagenic and carcinogenic potential. Since then, the Ames test has
been put into widespread use all over the world and thousands of
chemicals have been tested to evaluate their mutagenic and
carcinogenic potential. Many additional short-term tests were
developed to detect other types of genetic damage, as well as
nongenotoxic damage that would lead to cancer. With the advent of in
vitro metabolic activation, it seemed certain that the metabolism of
the whole animal could be mimicked on the Petri plate or in the test
tube and that whole animal assays would no longer be necessary.

As a result of large international collaborative studies to
evaluate various short-term in vitro and in vivo tests, it has become
clear that both approaches are needed and that any battery of assays
designed to evaluate environmental chemicals must include both in
vitro and in vivo short-term tests. The use of the Ames assay in
isolation has resulted in premature, and sometimes false, indictment
of potentially useful chemicals. A data base developed with the use
of this test alone is inadequate for the evaluation of a chemical's
mutagenic and carcinogenic potential in laboratory animals and
humans.

In this paper, I intend to review the results of three
international collaborative studies to evaluate the general utility

of short-term tests for mutagenicity and carcinogenicity and the
impact of the data base developed in these studies on the design of
testing schemes used in the safety evaluation of pesticides and
environmental chemicals in general.

International Collaborative Study to Evaluate Short-Term in Vitro Tests for Carcinogens

Two international collaborative studies (sponsored by the
International Program on Chemical Safety [IPCS] a collaborative
program of the World Health Organization, the International Labor
Organization and the United Nations Environmental Program) to
evaluate short-term tests for genotoxicity have recently been
completed (1,2). The first IPCS study was designed to evaluate
eukaryotic in vitro assay systems for use along with the Salmonella
reverse-mutation assay system. The Salmonella assay was shown to be
particularly suitable for use in screening environmental chemicals
for potential mutagenic and carcinogenic activity in the
International Program To Evaluate Short-Term Tests for Carcinogens
(IPESTTC) that was started in 1977 (3). In that study it was clear
that the Ames Salmonella test could not detect all known carcinogens
and it had to be used along with some other short-term test as yet to
be identified. In the IPCS in vitro study a total of 10 chemicals,
consisting of 8 carcinogens that are difficult to detect with the
Salmonella assay and 2 noncarcinogens, were used to evaluate the
utility of a wide variety of eukaryotic assays for genotoxicity. The
10 chemicals tested are as follows: carcinogens (acrylonitrile,
benzene, diethylhexylphthalate, diethylstilboestrol,
hexamethylphosphoramide, phenobarbitone, safrole and o-toluidine) and
noncarcinogens (benzoin and caprolactam). The eukaryotic systems
tested included; assays for gene mutation, gene conversion,
crossing-over and aneuploidy in fungi; assays for somatic cell
recombination and gene mutation in Drosophila; and assays for
metabolic cooperation, transformation, single-strand breaks,
unscheduled DNA systhesis, chromosome aberrations, sister-chromatid
exchange, micronucleus, polyploidy, aneuploidy and gene mutation in
mammalian cells in culture. An evaluation of the data base developed
in this study showed that the best overall performance was given by
the fungal assay for aneuploidy and the chromosome aberration assay
in mammalian cells in culture. Since the aneuploidy assay has had
limited use outside of the laboratory contributing test data to the
collaborative study, a recommendation for more general use will have
to await studies to evaluate interlaboratory reproducibility. Assays
for chromosome aberrations in mammalian cells in culture, which are
in widespread use, were suggested to complement the Salmonella
reverse-mutation assay (1). It is important to note, however, that
this recommendation was based on a study limited to only ten test
chemicals.
      In principle, justification for including tests complementary to
the Ames test is based on knowledge that genotoxic agents cause
damage through various mechanisms that may not be detectable in the
Ames test alone. For example, neither structural damage to
chromosomes nor the induction of aneuploidy can be detected in the
Ames test. As a result of such considerations, it follows that a
battery of short-term in vitro tests is required to provide a

comprehensive evaluation of the spectrum of genetic damage induced by
a genotoxic chemical.

In the initial reports on the utility of the Salmonella assays
for identifying carcinogenic potential (4-6) a high percentage of
known chemical carcinogens and a low percentage of known chemical
noncarcinogens were shown to give a positive response. One of the
problems encountered in IPESTTC (3) was the high frequency of
positive responses observed in vitro for 7 of the 14 chemicals
classified as noncarcinogens. The number of in vivo short-term tests
in IPESTTC was somewhat limited, but the total data base developed in
this experiment suggested that, although some noncarcinogens were
positive in vitro they were negative in in vivo short-term tests.
Two good examples of these differences were provided by the
noncarcinogens in the benzo(a)pyrene and pyrene (BP/P) and the
2-acetylaminofluorene and 4-acetylaminofluorene (2AAF/4AAF) pairs as
indicated in Table I.

Table I. Results of Assays on the Two Pairs of Chemicals BP/P and
2AAF/4AAF Reported in IPESTTC

Type of	Number of + Responses/Total Number of Assays			
Assay	BP	P	2AAF	4AAF
in vitro	40/47	17/40	32/36	27/35
in vivo	5/5	0/5	2/2	0/2

Thus, the rationale for the second IPCS study on in vivo
short-term tests was derived from IPESTTC and it was decided to
evaluate the ability of an even broader range of in vivo short-term
tests to determine which of these show the best descrimination
between known chemical carcinogens and noncarcinogens.

International Collaborative Study to Evaluate Short-Term in Vivo
Tests for Carcinogens

The second IPCS study on in vivo assay systems utilized the two
test chemical pairs BP/P and 2AAF/4AAF to evaluate a wide range of
whole animal short-term tests for genotoxicity and carcinogenicity.
In IPESTTC all 4 test chemicals gave positive results in a wide range
of in vitro short-term tests, and the objective of this second IPCS
study was to determine which in vivo assays were capable of
distinquishing the 2 carcinogens from the two noncarcinogens. The
results of this study have only recently been compiled in terms of a
final report (2), but the data show clearly that, in general, the
collective body of in vivo assays does discriminate between the two
carcinogens and the two noncarcinogens as shown in Table II.

Table II. Results of Assays on the Two Pairs of Chemicals BP/P and
           2AAF/4AAF Reported in the IPCS Collaborative Study on
           in vivo Assays

Type of Assay	Number of + and +/- Responses/Total Number of Assays			
	BP	P	2AAF	4AAF
in vivo	64/72	2/73	54/75	10/75

Particularly useful in vivo assays were the mouse bone marrow
micronucleus assay and the assay for unscheduled DNA synthesis in
cultured rat liver cells. Both assays showed good interlaboratory
reproducibility and they provide a reasonable battery of in vivo
short-term tests for further evaluation of in vitro genotoxins.

Effective Deployment of in Vitro and in Vivo Short-Term Tests for
Carcinogenicity

The effective deployment of short-term tests for individual chemicals
or their use in mass-screening programs has been the subject of
considerable debate in the recent literature (7). Obviously, if in
vitro genotoxins can be negative in vivo, a sequential scheme of
testing may be required to develop a data base that will permit a
comprehensive evalution of mutagenic and carcinogenic potential. One
possible approach to this problem is given in Figure 1.
     In this scheme a chemical is subjected to an in vitro battery of
tests (the Salmonella assay and an assay for chromosome aberrations
in mammalian cells in culture) to determine whether it is a genotoxin
in vitro and produces the type of genetic damage that can be detected
by each of these assays. A positive result in one or both of the
assays classifies the chemical as a genotoxin and the chemical is
then tested with a battery of in vivo assays. If both of the in
vitro assays are negative, then the chemical is tested in the
appropriate battery of assays to detect nongenotoxic carcinogens.
These tests may include assays for in vitro transformation,
unscheduled DNA systhesis, etc. Positive results with any of these
assays will classify the chemical as a potential carcinogen.
     In vitro genotoxins are then subjected to a battery of in vivo
tests (e.g. the mouse bone marrow micronucleus test and the rat liver
assay for unscheduled DNA synthesis). Since enzymatic detoxification
may well be organ-specific, one of the main problems in the in vivo
assays is the development of a group of tests that will permit a
comprehensive evaluation of the activity of the chemical being tested
in the appropriate target organs(liver, lung, kidney, bone marrow,
etc) as well as the gonads (ovary and testis). The dominant lethal
test in the rat has been used extensively to evaluate the mutagenic
effects of chemicals on germ cells, but positive results do not
always indicate genotoxicity (8). The data base on other germ cell
assays is much more limited and, in general, these tests involving
the use of rodent assays for genotoxicity are too costly for general
use (9). However, the data base on a new assay for unscheduled DNA
synthesis in the testis developed by Sega and presented as an

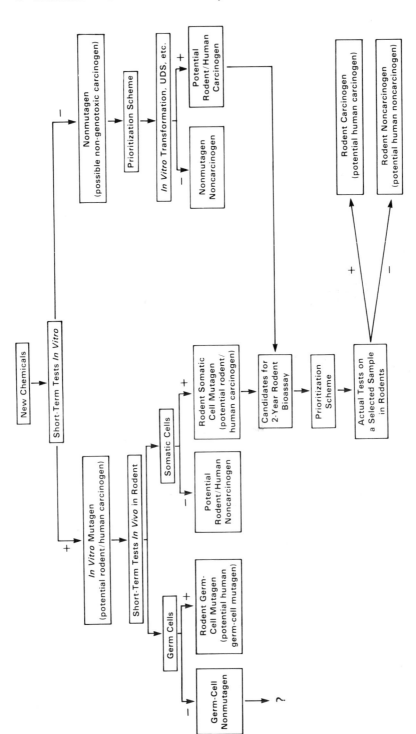

Figure 1. Proposed testing scheme for identification and classification of test chemicals with short-term in vitro and in vivo assays.

appendix to the Russell paper (9), looks promising. The development
of a sufficiently comprehensive battery of in vivo assays will
eventually permit the classification of those chemicals that give
negative results as potential rodent/human nongenotoxins and as
germ-cell nonmutagens. Chemicals that give positive results in one or
more assay will be classified as potential rodent/human carcinogens
and will have to be tested in the traditional two-year rodent
bioassay for cancer.

In summary, these studies have demonstrated the general utility
of short-term tests to evaulate the mutagenic and carcinogenic
potential of environmental chemicals. It is now clear that to
provide a comprehensive evaluation, it will be necessary to
supplement such short-term in vitro asssays as the Ames Salmonella
test and an in vitro assay for chromosome aberrations in mammalian
cells in culture with short-term in vivo assays. At present, the
best candidates for these in vivo assays are (1) the mouse bone
marrow micronucleus assay, and (2) the assay for unscheduled DNA
synthesis in rat liver cells.

Literature Cited

1. Ashby, J.; de Serres, F. J.; Draper, M.; Ishidate, M.,Jr.;
   Matter, B. E.; Shelby, M. D., Eds. "Evaluation of Short-term
   Tests for Carcinogens: Report on the International Programme on
   Chemical Safety's Collaborative Study on in vitro Assays";
   Progress in Mutation Research, Vol. 5, Elsevier: North Holland,
   Amsterdam, 1985.
2. Ashby, J.; de Serres, F. J.; Shelby, M. D.; Margolin, B. H.;
   Ishidate, M., Jr.; Becking, G. C., Eds. "Evaluation of Short-term
   Tests for Carcinogens: Report on the International Programme on
   Chemical Safety's Collaborative Study on in vivo Assays";
   Cambridge University Press: Cambridge (in press).
3. de Serres, F. J.; Ashby, J., Eds. "Evaluation of Short-term Tests
   for Carcinogens: Report of the International Collaborative
   Study"; Progress in Mutation Research, Vol. 1, Elsevier: North
   Holland, Amsterdam, 1981.
4. McCann, J.; Choi, E.; Yamasaki, E.; Ames, B. N. Proc. Natl. Acad.
   Sci.USA 1975, 72, 5135-9.
5. McCann, J.; Ames, B. N. Proc. Natl. Acad. Sci. USA 1976, 73,
   950-4.
6. Sugimura, T.; Yahagi, T.; Nagao, M.; Takeuchi, M.; Kawachi, T.;
   Hara, K.; Yamasaki, E.; Matsushima, T.; Hashimoto, Y.; Okada, M.
   IARC Scientific Publ. No. 12, International Agency for Research
   on Cancer; Lyon, 1976, pp. 81-101.
7. Ashby, J.; Purchase, I. F. H. Environ. Mutagen. 1985, 5,
   747-758.
8. Green, S.; Auletta, A.; Fabricant, J.; Kapp, R.; Manandhar, M.;
   Sheu, C.; Springer, J.; Whitfield, B. Mutat. Res. 1985, 154,
   49-67.
9. Russell, L. B.; Aaron, C. S.; de Serres, F.; Generoso, W. M.;
   Kannan, K. L.; Shelby, M.; Springer, J.; Voytek, P. Mutat. Res.
   1984, 134, 143-157.

RECEIVED October 7, 1986

Chapter 5

Simulation Modeling in Toxicology

J. T. Stevens and D. D. Sumner

Ciba-Geigy Corporation, P.O. Box 18300, Greensboro, NC 27419

Attempts to understand and manage toxicological manifestations are generally a reactive rather than a
predictive endeavor. Although we have responded by
addressing untoward reactions with no effect levels
and safety factors and oncogenic responses with quantitative and qualitative risk modeling, little has
been established as a foundation for prediction of
responses. The purpose of this paper will be to
present a summary of the state-of-the-art on structure activity modeling; this process may assist in
the evolution of predictive approaches to toxicology.

The performance of mechanistic studies to determine how xenobiotics produce their toxic responses is a promising approach for the
understanding of risks (1). This technique would appear equally
appropriate for a proactive, as well as reactive, examination of
the factors involved. Understanding is based on experience; this
commodity is as invaluable in proactive reasoning as reactive
scrutiny. As we explore approaches to Structure-Activity-
Relationship and biotransformation kinetics, it will become apparent that the foundation for the future is firmly anchored to the
past. The link between similarity of structure and similarity of
biological response is the key to making predictions on biological
and/or toxicological properties. Our ability to simulate in
models is only as good as our ability to accurately establish that
link.
    For this consideration of simulation modeling, some of the
currently available approaches for prediction of toxicity through
chemistry will be examined--looking at both integrated knowledge
and empirical evaluation.

Structure Activity Relationships

The information base between toxicological response and chemical
structures has been growing exponentially. The integration of
these data into a comprehensive and reliable structure activity

relationship system (SAR) has, generally, been linear. Although
computers have assisted greatly in this evolution, development has
required that certain basic criteria for effective assessment be
met. These requirements are included in Table I.

Table I. Criteria for Reliable Structure Activity Modeling

· Broad data base (>60 compounds)
 	
   - Congeneric chemistry (good prediction only within the
             same class of chemistry)
 	
   - Same mode/mechanism of action
 	
· Relevant test/organism

The incorporation of these criteria into a structure activity model
is no simple task. Although it is not necessary to have either
congeneric chemistry or the same mode of action for practical SAR
models, these data provide a foundation for greater accuracy. A
wide variety of biological reactions must be considered.
     Quantitative structure activity relationships (QSAR) have been
used in designing structures for efficacy for both pharmaceuticals
and agricultural chemicals. Hansch (2) was one of the first to
attempt to integrate the congeneric chemistry, mathematically, with
biological activity; a generalized Hansch regression equation resem-
bles the following (3):

$$\text{Log} \frac{1}{\text{Effective Dose}} = \frac{\text{Constant} + (\text{coefficient}_1 \times \text{parameter}_1) +}{(\text{coeff.}_2 \times \text{par.}_2) + \ldots + (\text{coeff.}_n \times \text{par.}_n)}$$

     Hansch incorporated several chemical/physical chemical charac-
teristics into this approach (4). He found Log P values (log
octanol/water portion coefficient) were usually applicable with
other parameters, such as Hammet linear free-energy relationships
and Van der Waals radii selectively applicable. Continued work in
this area by Hansch and other workers (5) has expanded the number of
relevant characteristics to include molecular orbital calculations
and diffusion parameters. Still, this quantitative approach embodies continuous parameters as an endpoint, a parametric philosophy.
     On the other hand, several investigators (6, 7) have taken
another approach, based on pattern recognition. These dichotomous
models search for agreement between dependent variables; i.e.,
whether a chemical entity or substructure can be associated with a
particular toxic property. For example, certain N-nitrosamine
groups are associated with tumors in animals. Since this consideration is not dependent on a relationship between the endpoint and the
dose, the quantitative term is dropped from QSAR and the effort
simply named SAR. This approach is best expressed by the dependent
equation:

Toxic Response = f (Chemical Structure)

This is an overwhelming concept when one envisions the number of permutations possible. The Food and Drug Administration (FDA) has tackled this formidable task with just 33 questions to define three presumptive toxicity categories (8). The categories are used to determine the degree of testing required by the FDA. This scheme, though basic and very pragmatic, represents a potentially useful tool for toxicologists, biologists and synthesis chemists. It offers a mechanism to recognize the potential hazard of a compound or to change the molecule to avoid unnecessary toxicity.

The 33 questions on structure are answered by a yes or no (Appendix 1). Each answer leads either to a further question or to the classification into one of three classes of presumptive toxicity (Table II).

Table II. FDA Presumptive Toxicity Classifications

Description	Classification
Relatively innocuous	I
Intermediately toxic	II
Presumptively mostly toxic	III

The requirements for this approach are simple. The structural formula, as well as a knowledge of chemistry and biology are used to make judgements on metabolism.

The FDA has evaluated the reliability of the scheme, using literature data and FDA's inventory of over 1,500 substances and the no observable effect levels derived from subchronic and chronic studies. Thus far, in a retrospective and prospective review, based upon the available information at the Agency, the FDA has no indication that these 33 questions do not adequately classify compounds (cross-comparison of structure to findings) into the three presumptive classes.

However, it is obvious that the FDA approach is a generalized toxicity classification and cannot supply the answers to questions such as, what are the metabolites and which compounds will be teratogens, mutagens or oncogens. Although the FDA approach has built its foundation on a broad data base, it does not narrow its spectrum to a precise toxicological response or mode of action.

It has been clear for some time that pattern recognition approaches would not go far without the computer (9). Utilizing techniques, such as regression, discriminant, and factor analyses progress in SAR (10) has been further enhanced. This evolution has led to such useful tools as described in Table III.

Table III.   Computer Assisted Research Models

Model	Function
• Computer Assisted Synthesis Planning (CASP)	• Can predict precursor's need for synthesis as well as most likely metabolites
• Constrained Structure Generation (CONGEN)	• Can generate all possible isomers if the substructure elements and the molecular formula are provided
• ADAPT (11)	• Affords an opportunity to predict the activity and properties of unstudied structures through application of pattern recognition and statistics

Although to the synthesis chemist, CASP and CONGEN may seem highly intriguing, to the biologist, a system such as ADAPT opens the door to the design of new and efficacious molecules for a myriad of uses. In fact, many of the major chemical industries have begun to incorporate such computer assisted systems into their research programs as a component of informed design, rather than the formerly predominant serendipitous discovery. These SAR techniques have not supplanted standard biological efficacy models; however, the information gained helps to establish the foundation for enhanced pattern recognition.

In pattern recognition modeling, such as ADAPT, it is difficult to effectively visualize and manipulate chemical structure. Instead, there has been an effort to translate abstract structure into quantities and/or numerical entities (10), referred to as molecular descriptors. Such descriptors have been classified as presented in Table IV.

Table IV.   Molecular Descriptors Used in SAR

Classes of Descriptors	Examples
• Geometric/Biophysical	Rotation axes, molecular volume and surface area
• Physiochemical	Log P, atomic charges, linear-free energy relationships
• Structural	Molecular weight, atomic numbers, types of bonding, molecular orbital calculations, ring structures
• Substructural/ topological	Topological and physiochemical properties of substructural arrangements, molecular symmetry and/or bonding

It is obviously not possible to unravel the entire complexity of the physical, chemical and biological properties of even the simplest of molecules. However, focusing on the apparently pertinent descriptors for structures, one can, via pattern recognition, begin to equate toxicological response with structures:

Toxicological response = f(structure) = f(molecular descriptors)

Jurs et al (10) qualitatively and quantitatively examined the correlation of a variety of molecular descriptors for polycyclic aromatic and nitroamine compounds with carcinogenesis. The result of these efforts has been the evolution of predictive equations which capture the oncogenic response for these classes of compounds as a function of the molecular descriptors.

Enslein and coworkers (12, 13, 14, 15) have utilized this approach to develop predictive models for carcinogenicity, teratogenicity and mutagenicity, as well as for acute toxicity endpoints.

This approach in predictive toxicology has manifest itself by the incorporation of certain key principles. These include:
· Marker compounds, compounds with a known biological endpoint, used to produce predictive equations.
· Equations are used for comparison of unknown compounds and to test the system.
· Finally, a statistical approach, such as stepwise regression (if endpoint is continuous) or discriminant analyses (if the endpoint is categorical) to verify the quality of fit.

Despite the great success that has been achieved with the approach taken by Enslein and coworkers, the utility of the system is limited by the depth of the database available in the open literature. If proprietary data were available from the files of pharmaceutical and agricultural companies, a new dimension to the reliability might be added. The possibility that new compounds can be examined for toxicity before they are synthesized is intriguing. However, the release of proprietary information from the bulwark of inherently competitive organizations is not likely in the near future. Therefore, Dr. Enslein plans to make his software available by mid-1987. Then, perhaps, the criteria for reliable structure activity modeling in the area of toxicology may be better served. However, until these criteria are achieved, it will be essential to rely on the more pragmatic approaches to simulation modeling; that is, bioassays.

Predictive Empirical Systems

With pressures from the animal rights movement, an impetus has been generated for the development of in vitro and/or computer models to reduce the level of in vivo testing. In the seventies, the hope of the future was placed in what was then considered a potential replacement technique for the lifetime rodent bioassay for cancer assessment--the short-term mutagenicity tests, particularly the Ames Evaluation (16). Brusick (17) has shown that the correlation between a positive mouse bioassay and a positive rat bioassay for a selected group of materials is no better than the

match for a positive Ames and/or a positive rat or mouse bioassay
(Table V). In addition, a comparison of the Enslein's SAR Carcinogen model (18) for these same human carcinogens is provided.

Table V.   Chemicals Evaluated as Positive for Carcinogenicity
           in Humans Compared to the Response in Rodent and
                Bacterial Predictive Assays, D. Brusick (17)

Chemical	Rat Bioassay	Mouse Bioassay	Ames Test	Enslein's Model
4-Aminobiphenyl	+	+	+	+
Arsenic	−	−	−	0
Asbestos	+	+	−	0
Benzene	−	−	−	0
Benzidine	+	+	+	+
Bis(chloromethyl)ether	+	+	+	0
Chromium and some chromium compounds	+	−	+	0
Cyclophosphamide	+	+	+	+
Diethylstilbestrol	+	+	−	0
Melphalan	+.	+	+	+
Mustard Gas	No Data	+	+	+
2-Naphthylamine	−	+	+	+
Soot, tars	−	+	+	0
Vinyl chloride	+	+	+	+
Percent predictability of humans	69	79	71	100
+ = positive − = negative 0 = cannot be evaluated				

The data from Enslein's Model show that a good match with the
rodent bioassays is possible for organics upon which the SAR model
is based. The SAR model is ineffective for metals, simple hydrocarbons like benzene, mixtures and hormones. Within these limits,
the prediction is 100%. Considering the reliability of rodent and
in vitro bioassays to predict the human response, it is possible,
with continued development, that in the future, SAR may become a
powerful tool to supplement our other sources of toxicological
information.

Appendix 1

		If 'No'	If 'Yes'
		...Proceed to...	
1.	Normal body constituent?	2	I
2.	Certain nitrogen FG's?	3	III
3.	"Non-physiological" elements?	5	4
4.	Innocuous salt of above?	III	7
5.	Simple ... HC or common CHO?	6	I
6.	Certain p-alkoxy benzenes?	7	III
7.	Heterocyclic?	16	8
8.	Lactone or cyclic diester?	10, 20,10	9
9.	Certain lactones?	23	III
10.	Three-member heterocycle?	11	III
11.	Hetero ring; strange FG's ...?	12	33
12.	Heteroaromatic?	22	13
13.	Any substituents?	III	14
14.	More than one aromatic ring?	22	15
15.	Readily hydrolyzed ...?	33	22(16)
16.	Common terpene?	17	I
17.	Readily hydrolyzed ...?	19	18
18.	Is it one of ------?	I	II
19.	Open chain?	23	20
20.	Linear or simply branched aliphatic ...?	22	21

		If 'No'	If 'Yes'
		...Proceed to...	
21.	Three or more types of FG's?	18	III
22.	Common component of food or structurally closely related ...?	33	II
23.	Aromatic?	24	27
24.	Monocarbocyclic; certain FG's ...?	25	18
25.	Cyclopropane or cyclobutane ...?	26	II
26.	No unusual FG's; certain ketones?	22	II
27.	Any ring substituents?	III	28
28.	More than one aromatic ring?	30	29
29.	Readily hydrolyzed ...?	33	30,18
30.	Other than certain substituents?	18(19)	31
31.	Acyclic acetal, -ketal, -ester ...?	32	18,19
32.	Only certain FG's, plus ...?	22	II
33.	Enough sulfonate/sulfamate?	III	I

Literature Cited

1. Stevens, J. T.; Sumner, D. D. J. Toxicol.-Clin. Toxicol. 1983, 19,781.
2. Hansch, C. Chem. Res. 1969, 2, 232.
3. Enslein, K. Pharmacol. Rev. 1984, 36, 131S.
4. Leo, A; Jow, P.Y.C.; Silipo, C.; Hansch, C. J. Med. Chem. 1975, 18, 865.
5. "QSAR and Strategies in the Design of Bioactive Compounds," Seydel, J. K., ed. Verlagsgesellschaft, Weinheim, FRG. 1985.
6. Jurs, P. C.; Chou, J. T.; Yuan, M. J. Med. Chem. 1979, 22, 476.
7. Tinker, J. F. J. Computational Chemistry. 1981, 2, 231.
8. Rulis, A. The Toxicology Forum; 1982 Annual Summer Mtg. 1982, p. 352.
9. Jurs, P. C.; Ham, C. L.; Brugger, W. E. In "Odor Quality and Chemical Structure"; Moskowitz, H. R.; Warren, C. B; Eds.; ACS Symposium Series No. 148, American Chemical Society: Washington, D.C., 1981; pp. 143-160.

10. Jurs, P. C.; Hasan, M. N.; Henry, D. R.; Stouch, T. R.; Whalen-Pederson, E. K. Fund. Appl. Toxicol. 1983, 3, 343.
11. Stuper, A. J.; Brugger, W. E.; Jurs, P. C. In "Computer Assisted Studies of Chemical Structure and Biological Function;" Wiley-Interscience, New York, 1979.
12. Enslein, K.; Craig, P. N. J. Toxicol. Environ. Health 1982, 10, 521.
13. Enslein, K.; Lander, T. R.; Strange, J. R. Teratog. Carcinog. Mutagen. 1983, 3, 289.
14. Enslein, K.; Lander, T. R.; Tomb, M. E.; Landis, W. G. Teratog. Carcinog. Mutagen. 1984, 6
15. Enslein, K.; Lander, T. R.; Tomb, M. E.; Craig, P. N. Benchmark Papers Toxicol. 1983, 1, 1.
16. Wolff, G. L.; J. Environ. Pathol. Toxicol. 1977,1,79
17. Brusick, D. In "Application of Biological Markers to Carcinogenicity Testing"; Milman, H. A.; Sell, S.; Eds. Plenum Press, N.Y., 1983; pp. 153-163.
18. Enslein, K., personal communication.

RECEIVED October 14, 1986

PESTS

Chapter 6

Vulnerability of Pests: Study and Exploitation for Safer Chemical Control

Robert M. Hollingworth

Department of Entomology, Purdue University, West Lafayette, IN 47907

Despite considerable improvements in their safety, the
current use of pesticides causes an uncertain but
disturbing level of toxicity to non-target organisms on
a worldwide basis. A better knowledge of the
biochemistry and physiology of pests could reduce this
threat by decreasing unintended exposure to existing
pesticides and by aiding in the discovery of new and
more selective ones. Such knowledge can help in the
discovery process in several ways -- in the development
of new control concepts, in the rational design of
novel compounds, and by providing tools for their
efficient evaluation and optimization. Scientifically-
based strategies to slow the onset of resistance to
such rare and valuable materials must be developed.
Here, too, a better knowledge of their biochemical
modes of action and of pest vulnerability and defenses
will be indispensible. However these goals can be fully
realized only if there is greater investment in research
into pesticidal mechanisms and responses in target and
non-target species.

For the last 25 years and more the use of pesticides has been a
controversial and troubled subject of continuing public concern. This
concern has been based on the feeling that pesticides, as presently
used, may be seriously hazardous to man and to the environment. The
degree to which this is true is not clear. On the one hand reasonably
complete statistics indicate that in the USA, which uses about 30% of
the pesticide produced in the world, less than 50 people a year are
killed by accidental exposure to these materials (1), fewer than the
number killed by lightning or by insect stings. Statistics for other
developed nations show similar mortality rates. However, worldwide
estimates of the number of accidental deaths from pesticides, though
very uncertain, are much higher, ranging from 5,000 to over 20,000
per year in the early 1970's (2), and indicating a considerable
degree of misuse in developing nations. Nor has this situation
clearly improved in the last decade. About 13,000 pesticide-related

0097-6156/87/0336-0054$06.75/0
© 1987 American Chemical Society

hospitalizations and 1000 deaths (73% as suicides) were recorded in
the island of Sri Lanka each year from 1975 to 1980 (3), and 500-650
such deaths were recorded in the Phillipines in 1980-1981. The
herbicide paraquat accounted for 93 known fatalities in the small
island of Trinidad in the single year of 1984 (4).

Estimates of the number of people injured by pesticides vary
widely but may be 100-times the accidental death rate on a worldwide
basis (5,6). In 1985, 2500 occupationally-related accidental
pesticide poisonings were officially reported in California and the
real number of injuries may be considerably higher. Neither is the
record of safety in pesticide manufacturing in the USA unblemished.
Serious injuries to workers involved in manufacturing the insecticide
chlordecone, the nematicide dibromochloropropane (DBCP), and,
possibly, the insecticide leptophos have been reported in the last 15
years (4,5). There is preliminary but disquieting epidemiological
evidence that frequent users of some herbicides may have an elevated
risk of contracting cancer, particularly if standard safety
precautions are not observed (e.g. 7,8).

The toll on wildlife is uncertain, but the role of DDT and other
organochlorine residues in the decline of some raptorial bird
populations is reasonably well established and there are regular
reports of the death of birds due to exposure to organophosphate
insecticides such as diazinon and famphur during normal use, among
many other incidents and concerns (9).

It is estimated that at least 50% of food items in the USA
contain detectable pesticide residues (6). A considerable segment
(77% in a recent poll) of the US population is concerned about
pesticide residues in food, and, increasingly, about residues in
groundwater and other water that is used for drinking, even though
the toxicological significance of such low level exposure is dubious
and the results may be mainly a tribute to the sensitivity of modern
analytical equipment. These and parallel examples of exposure through
spray drift lead to concern and are poorly tolerated by those
involved because of their involuntary nature, despite (or perhaps
because of) the unclear nature of the degree and type of hazard
involved.

Yet pesticides are regarded as essential for modern society --
as a necessary economic investment in agriculture and as a vital tool
in the control of vector-borne diseases such as malaria and
onchocerciasis. This leads to a tense and adversarial situation
between proponents and opponents of pesticides. The situation is
appropriately characterized as "you can't live with them and you
can't live without them".

In part, the increased pesticidal potency of newer compounds is
helping to solve some of these problems. Use rates of such compounds
as the photostable pyrethroid and avermectin insecticide/acaricides,
the sulfonylurea herbicides, and ergosterol biosynthesis inhibitors
as fungicides are a few ounces of active ingredient per acre at most.
This greatly reduces residues and decreases the chance of accidental
poisonings. However, these compounds are not entirely without hazard.
Many of the pyrethroids are exquisitely toxic to aquatic species,
including fish, and some avermectins have an acute oral toxicity to
rats considerably higher than that of parathion. The problem, even in
these examples of modern highly potent insecticides is that they act
on biochemical target sites which are also present and critical in

vertebrates i.e. the sodium channel of excitable tissues in the case
of pyrethroids and GABA-ergic inhibitory neurotransmission in the
case of avermectins.
Insecticides as a class present the greatest risk of acute
poisoning in man and other vertebrates. They are the most difficult
class for the discovery of new and safer materials. At the same time
the usefulness of the limited range of existing compounds is
continually threatened by the development of resistance.
Consequently, insecticides represent both the biggest need and the
greatest challenge in devising safer pest control technologies, and
special emphasis is laid on insect control in the discussion that
follows.
This article addresses the question of how we may increase the
safety of pesticide use through a knowledge of pest biochemistry and
physiology. Such research and knowledge can lead to improved safety
both in terms of compounds currently registered for use and, more
particularly, in helping guide the search for new and safer materials
to replace them. In this, it deals with possibilities only. Which, if
any, of these possibilites will come to fruition depends on many
factors including the level of investment in their development and
potentially severe technical, economic, and regulatory constraints.

Enhancing the Safety of Existing Compounds

The risk from pesticide use depends on two factors — toxicity and
exposure. Since the toxicity of an existing compound is an innate
property, it cannot be altered, but the toxicity of the final
formulation may be improved and this will alter the risk. To make the
use of current compounds safer, then, is mainly to concentrate on the
reduction of exposure or the improvement of formulations and
application methodologies for greater safety. Knowledge of pest
biochemistry and physiology could help to achieve this in several
ways.

Minimizing exposure through reduced application rates or frequency.
Attractants and baits for mobile pests may greatly safen the use of
existing compounds by allowing the localized placement of an
insecticide or by decreasing the overall concentration needed in the
formulation. An excellent example is the use of bait formulations of
the insecticide, mirex, which reduced the amount of this
organochlorine material needed to control imported fire ants from
about 1 kg to about 1 g per acre (10). Unfortunately, even this
reduction was not judged to be enough to allow the continued use of
mirex for this purpose because of its environmental stability and
potential carcinogenicity. Also, the use of insect pheromones to
monitor insect populations and thereby reduce the frequency and
improve the timing of pesticide applications is well established
(11). Applications may be reduced by as much as 50% and this in turn
increases the contribution to control by beneficial insects, further
decreasing the need for insecticide treatment. In some favorable
cases, pheromones can be used to attract insects to localized
insecticide-treated sites for safer control. It is salutory to
remember that early basic research on insect sex pheromones was
essential for these benefits, but was derided in some quarters as a
waste of public funds on studying the sex lives of insects.

It is probable that through a better understanding of such
examples of chemoreception and the feeding behavior of pests, new,
improved attractants and bait formulations could be produced.
Particularly important in this regard are such insects as the
Heliothis complex on cotton which are exposed to foliar insecticides
only briefly as they hatch from the egg and migrate to the cotton
boll. Much insecticide is wasted in trying to ensure that a lethal
dose is picked up by contact during this short time. The brief window
of vulnerability also imposes a severe limitation on the efficacy of
such highly safe materials as the Bacillus thuringiensis endotoxin,
which must be ingested by the larvae to be toxic.

Compounds that alter insect behavior in other ways may also be
useful e.g. chemicals that increase insect locomotor activity should
enhance the uptake of pesticides from treated surfaces. Such
compounds are already known among the formamidines insecticides and
their relatives that stimulate octopamine receptors ([12]), and further
research in this area of pest biochemistry could reveal other
locomotor stimulators such as phosphodiesterase inhibitors ([13,14]). A
similar philosophy underlies recent studies showing that alarm
pheromones increase the motility of aphids and that this results in
the enhanced efficacy of contact insecticides such as pyrethroids and
organophosphates ([15]).

It is also probable that a more detailed study of the factors
governing the uptake of pesticides from treated surfaces could lead
to improved formulations of higher efficiency and lower concentration. From the limited data available it is clear that events on the
surface of the leaf or insect and the type of formulation applied may
radically affect the performance of herbicides ([16]) and insecticides
([17]).

Modification of the Toxicity of Pesticides with Formulation
Additives. Potentially the use of additives to modulate the
toxicity of pesticides could lead to a considerable increase in
their safety to non-target species, man included. This concept has
been well explored and exploited in adding safeners to certain
herbicides such as the thiocarbamates. These compounds stimulate
defensive metabolic reactions in the crop species but not in weeds
([18,19]). This principle has also been applied to vertebrates,
but only to a very limited degree. Under some circumstances the
thiocarbamate rice herbicide, molinate, may show toxicity to carp in
nearby ponds. Based on a knowledge of the safeners that are active in
plants, a compound was discovered that, when applied with molinate,
acted as an antidote/safener for the carp ([20]). Little
effort, either theoretical or empirical, seems to have gone into
developing other such examples.

Since "safening" involves transiently changing the biochemistry
of non-target species, the alternative strategy of making the
pesticide selectively more active with synergists that alter the
pest's biochemistry may be more appealing. Numerous examples of
synergism with all types of pesticides are known, but the present
applications of the principle are limited by several factors
including a lack of selectivity for non-target species in the
synergistic process ([19,21]). The discovery and application of
pest-specific synergists is a realistic goal which, if exploited,
could lead to considerably reduced formulation and application rates

and thus to a higher level of safety to the applicator and
environment. One compound for which such an approach is justified is
the herbicide paraquat which, based on the known number of human
poisonings, is one of the most hazardous pesticides currently in use.
In both plants and animals its toxic actions result from the
catalytic generation of reactive forms of oxygen in the tissues.
Success in the use of copper and zinc chelators to synergize its
toxicity by inhibiting the antioxidant enzymes (e.g. superoxide
dismutase) that tend to protect the plant is therefore promising
(22). Synergists that are relatively selective between plants and
animals in this regard would be particularly valuable. Such
synergists, selective or not, could not have been devised without a
thorough knowledge of the mechanism of action and physiological
responses to paraquat in plants. The concept of selective synergism
of pesticides has considerable potential but, as yet, has hardly been
exploited.

Strategies to Overcome Pesticide Resistance

The development of resistance is a process which inexorably
eliminates existing pesticides from the market, safe and dangerous
alike. Even if they are not eliminated, application rates may have
to be raised as resistance develops leading to enhanced levels of
exposure for non-target organisms. Even with herbicides, resistance
is now starting to become a practical problem; it is an established
one for insecticides and the modern selective fungicides. It is
probably a worse threat to safer compounds since these often are
selectively toxic by attacking a single enzyme or receptor peculiar
to the pest group. A single mutation at this site may then render
the target insensitive and the pest highly resistant. Compounds
acting at multiple essential sites are less open to the development
of target site resistance, but by the same token are less likely to
be selective in their toxic actions. Preserving safe pesticides by
slowing, preventing, or reversing the onset of resistance depends
absolutely on understanding the biochemistry and physiology of the
pest and the biochemical and population genetics of the resistance
process.
    After many years of relative passivity in the face of
resistance, a more aggressive concept of "resistance management" has
developed considerable momentum in the last decade as it has become
clear that new pesticides will not be readily available to replace
those lost to resistance. This approach is reviewed elsewhere much
more fully than is possible here (21,23-26). There are several
critical areas of resistance management where a knowledge of pest
biochemistry and physiology is essential:

1. Definition of existing and potential resistance mechanisms
    and their genetic basis.

2. Development of rapid, simple, and cheap methods to monitor
    resistance levels and mechanisms occurring at a low level in
    the field.

3. Development of chemical strategies to alleviate or prevent
    resistance.

Resistance to pesticides arises primarily through changes in the sensitivity of the site of action or in the metabolism of the pesticide (25,27,28). Many pesticides are activated metabolically. While it is theoretically possible to generate resistance through reduced activation, it seems much more common to observe increased detoxification in resistant strains. In some cases decreased uptake or enhanced excretion also contribute. It is an obvious prerequisite for any type of scientifically-based attempt to combat resistance that the resistance mechanism and its genetic basis must be defined.

There is a clear need for rapid, simple and cheap methods to monitor the status of resistance mechanisms and levels while they are still at a low incidence in field populations. If this can be achieved, early warning of changes in resistance gene frequency within the population will be possible. This should allow a shift to alternative control measures before the gene frequency and associated development of a supportive genotype progress to a stage which results in the irreversible loss of a desirable pesticide. The same survey methods will also be essential for monitoring the success of these alternative strategies. To be successful, alterations in gene frequency need to be detected when no more than a few percent of the population has the resistance gene. This demands a high degree of sensitivity and reliability at the population level that is not possible with typical bioassay methods. However biochemical tests for the presence of the resistance gene in individuals in the population may meet these requirements.

Some examples of approaches to such tests are already available e.g. assays for esterase levels in individual leaf- and planthoppers, mosquitoes, and aphids that could be run under field conditions or in local laboratories in the search for individuals with an elevated activity that provides resistance to organophosphates and other esters (29,30). Other simple tests for enzymological markers of resistance seem feasible based on the catalytic activity of the resistance site such as changes in the sensitivity of acetylcholinesterase (AChE) to inhibitors (29), mixed function oxidase (MFO) activity, and glutathione transferase activity. In the herbicide area, a field test for altered sensitivity to photosynthesis inhibitors has been described (31). However, developing specific diagnostic methods for some other traits involving changed target sites such as the kdr resistance mechanism for pyrethroids will be more challenging. With the purification of the changed enzymes or receptors responsible for resistance has come the possibility of gene cloning and the development of highly sensitive and specific diagnostic tests based on immunological methods such as ELISA assays (32,33).

The third element in resistance management is the development of strategies to alleviate resistance. Several possibilities can be envisioned (Table I). The various approaches in Table I and their difficulties have all been discussed previously (21,25,26), and some have been exploited on occasion by an essentially empirical approach. The intelligent use of mixtures or alternations of dissimilar materials could be viewed as an attempt to gain the resistance-delaying advantages of a pesticide with multiple sites of action. Although widely used in fungicidal treatments, the rational development of this approach is still in its infancy. Besides the desired suppression of the onset of resistance, it also carries the hazard of the rapid development of resistance to both types of pesticide, e.g. by the selection for a detoxification mechanism that affects both compounds. Predictions of which of these outcome may occur are still hazardous because of our lack of basic knowledge. All of the methods in Table I will be more accessible and predictable only as our understanding of pest biochemistry and physiology increases.

Table I. Some Chemical Strategies to Alleviate Resistance.

---

1. Use of pesticides with multiple sites of action.

2. Use of mixtures of compounds with dissimilar modes of action that lack cross-resistance potential.

3. Alternations and rotations of such dissimilar compounds.

4. Use of additives that antagonize the adaptive value of the resistance mechanism.

5. Use of compounds that display negatively-correlated cross-resistance.

In a number of instances the use of synergists to antagonize the adaptive advantage of a metabolic resistance mechanism has been tried, but with varying results (34). One successful example is the use of esterase inhibitors to prevent resistance to organophosphates in leaf- and planthoppers in Japan (35). Once the mechanism of resistance is understood at the biochemical level, strategies to combat other types of resistance can be envisioned e.g. the use of mitochondrial poisons to prevent the energy-dependent expulsion of sterol biosynthesis inhibitors by some fungi (36).

It is harder, but not impossible, to envision synergists that act to oppose a loss of sensitivity at the site of action. However, in these cases, other strategies are available since a biologically-essential target site (enzyme, receptor) generally cannot disappear but is only changed somewhat in its properties in the mutant form. This implies that in some cases, other compounds may be found that are effective against the changed site and which therefore are toxic to the resistant pest. This situation, which should be powerful in preventing or even reversing the effects of resistance, is an example of negatively-correlated cross-resistance i.e. the higher toxicity of a compound to the resistant than the susceptible strain. Such compounds have been been discovered and are already being utilized in a limited number of situations. For example, it has been found that resistance to typical carbamate insecticides in the green rice leafhopper often involves the development of a mutant form of AChE which is insensitive to inhibition by N-methylcarbamates. However, a site on the changed AChE is now sensitive to N-propylcarbamates which are relatively inactive on the native enzyme (37; Figure 1). Presumably this site has undergone a change in topography that allows

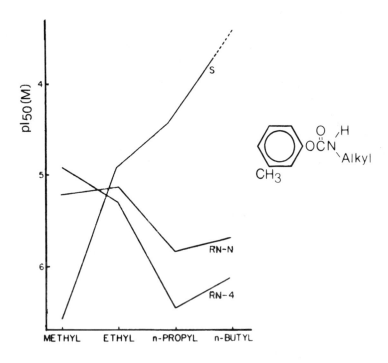

Figure 1. Comparative inhibition (pI50) of acetylcholinesterase from susceptible (S) and two resistant (RN-N and RN-4) strains of green rice leafhopper by m-tolyl N-alkylcarbamates in which the N-alkyl group varies from methyl to n-butyl. Reproduced with permission from Ref. 37. Copyright 1983 Plenum Press.

greater bulk in this N-alkyl part of the inhibitor molecule. The
difference in sensitivities of the forms of AChE to inhibition is
paralleled by differences in the toxicity of N-methyl and
N-propylcarbamates to susceptible and resistant leafhoppers (27).
More recently a comparable enhanced inhibition in resistant strains
has been observed with aryloxadiazolone anticholinesterases (38). A
second promising example is the discovery that some natural and
synthetic isobutylamides are selectively toxic against houseflies
that carry the super-kdr resistance trait (39). This gene causes an
alteration in the sensitivity of the site of action for DDT and
pyrethroids and is a major threat to the continued efficacy of
synthetic pyrethroids in many of their applications.
        Turning to the fungicide area, it has been shown that resistance
to benzimidazole fungicides often results from a change in the
benzimidazole binding site on beta-tubulin or from enhanced stability
of the microtubules. Some herbicidal N-phenylcarbamates which affect
microtubule functioning in plants have been found to be specifically
toxic to benzimidazole-resistant strains of fungi (28). Field tests
are currently being conducted with one analog that is not herbicidal,
S-32165 (diethofencarb; isopropyl 3,4-diethoxyphenylcarbamate), to
evaluate its use as a "resistance breaker" (40). Another example of
negatively-correlated cross-resistance among fungicides involving
phosphorothiolates and phosphoramidates has been described (41).
        As information on the molecular architecture and mechanisms of
pesticide sites of action becomes more generally available and the
nature and effect of mutations on their sensitivity to pesticides is
defined, it should become possible in some cases to design agents
that specifically interfere with the altered site of the resistant
forms. The same line of reasoning suggests that the design of
specific and selective synergists to block individual metabolic
resistance mechanisms may eventually be possible.

New and Safer Pesticides - Exploiting Pest Vulnerability

The defects of current pesticides, regulatory actions resulting
from them, the natural desire of chemical companies to discover new
and better compounds than their competitors, and the regular loss of
pesticide efficacy due to resistance and changes in agricultural
technology all provide impetus to the search for new compounds.
However, increased research and regulatory demands are expensive and
the difficulty of finding molecules with clearly superior properties
to those of existing compounds has made the discovery of new
compounds with improved safety characteristics an increasingly
costly and rare occurrence. It has been pointed out that the peak
of innovation in the introduction of new pesticides was seen in the
late 1960's at about 20 compounds per year. Since then it has
declined precipitously (25,42,43). Recently the success rate is
estimated to be one compound commercialized for every 15,000 to
20,000 synthesized with the total cost of developing this single new
compound being anywhere from $20 million (25) to $45 million (44).
Pesticide discovery is a game of roulette on a wheel with many
thousands of numbers. To stay in this game long enough to hit winners
and recoup investment is extremely expensive. An increasing number of
companies have decided that they cannot afford to play and have
cashed in their chips. In itself, this decreases our chances of
discovering safer pesticides.

On the basis of the old advice to know thy enemy, can studies of
pest physiology and biochemistry aid in the struggle to find new and
safer chemicals for pest control? The answer is by no means clear,
but it is obvious that the figures for success using past methods of
blind screening and analog synthesis are increasingly unfavorable and
uneconomical. The only alternative in sight is to attempt to apply
the growing knowledge of pest biochemistry and physiology to define
specific vulnerabilities in the pest and thus to decrease the odds in
our search for better materials. This prospect has been discussed
recently by a number of authors (19,20,43,45-48). Although the topic
of the genetic engineering as a means of pest control is beyond the
scope of this chapter, it equally depends on such a knowledge of the
enemy. Fortunately it is likely that research on pest biochemistry
and physiology can help improve our chances of finding novel
compounds, particularly safer ones, at several different levels.

Improved Bioevaluation and Optimization of Leads. At the simplest
level, such information provides improved methods for screening and
optimizing the activity in a new series of potential pesticides, and
to some extent in the discovery of such leads also.

Compounds with novel chemistry and marginal biological activity
appear more or less frequently in screening, and it is important to
decide whether these represent important leads for new products or
dead ends. Low activity may be due to any one of several factors
singly or in combination. Studies on pest physiology and
biochemistry provide a battery of tools to aid in making this
decision. Poor uptake can often be detected if the compound is
injected into the organism or applied in a special solvent such as
dimethyl sulfoxide, too-rapid metabolism is indicated by the
synergistic effect of co-application with specific enzyme
inhibitors, and low intrinsic activity can be detected by specific
assays of the action on the target site, when this is known. This
modest goal of integrating our present knowledge of the pest and its
defenses into the discovery and optimization process sounds simple
but even now is not always achieved.

A good example of how this process can work is provided by
the discovery of the nitromethylene heterocyclic (NMH) insecticides
at Shell Development Co., Modesto, CA. The initial NMH's were
obtained from an outside source as random screening items. In the
screen it was found that one NMH had 1% of the activity of parathion
against houseflies. This is not a very impressive level of activity,
but since the NMH's were new chemistry, there was a rapid follow up
in which the compound was injected into houseflies to subvert
penetration barriers and in the presence of the synergist, sesamex,
to inhibit oxidative degradation. This increased its activity against
the housefly to 50% of parathion's activity, which encouraged further
analog synthesis. At the same time, mode of action studies were
initiated. The strong excitatory symptoms in vivo suggested that the
nervous system was involved. The compound was applied to the 6th
abdominal ganglion preparation from the American cockroach where a
massive increase in nerve activity was observed followed by a block
of neurotransmission (Figure 2). These observations, supplemented by
further study based on the known properties of this preparation,
suggested that the NMH's acted to stimulate acetylcholine receptors

Compound 13

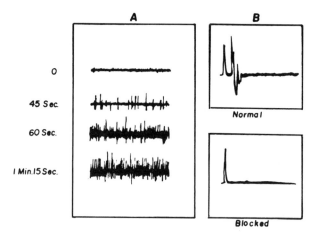

Figure 2. The development of spontaneous giant fiber discharges
(A) and the subsequent block of the response evoked by electrical
stimulation (B) in the central nervous system of the American
cockroach treated with a 3 μM solution of NMH compound 13
illustrated. Reproduced with permission from Ref. 49. Copyright
1984 Academic Press.

(49). This is a target site for very few current insecticides, and
resistance to compounds acting there has not been developed. This
discovery gave additional momentum to the synthesis program.
Eventually compounds with activity comparable to the synthetic
pyrethroids were discovered (50) and investigations in this area
continue.

The discovery of a mechanism of action is not always so readily
achieved, and may be very challenging if it is completely novel.
However, here pesticide resistance can offer some notable
advantages. Strains of pests with carefully defined single resistance
mechanisms involving insensitive sites of action, such as pyrethroid
resistant insects with the kdr gene or plants with an altered
triazine-binding site in the chloroplasts, can, by their relative
resistance to a new compound, be a useful and very rapid diagnostic
aid for evaluating its mode of action. Comparative studies with
strains having elevated levels of such defensive enzymes as MFO and
GSH transferase could help to define the susceptibility to metabolism
as a limiting factor in new compounds, and indicate the potential for
resistance to them by existing genetic mechanisms.

Rapid in vitro tests for actions at sensitive target sites, such
as enzyme and receptor binding assays, are needed to help in defining
modes of action early in the development of a new series, and in
guiding the chemist as changes in instrinsic activity occur during
its optimization. Currently we do have good assays for some enzymes
and for ACh receptors in insects, but methods for other sites need to
be developed for routine use. This is in strong contrast to the
pharmaceutical industry where in vitro tests are advanced and widely
used e.g. batteries of receptor binding assays to help define not
only the primary activity of a new compound but also its effects on
other receptor types as an indicator of potential side-effects.
Further basic research is essential to provide this capability with
pesticides. This may eventually allow us to use in vitro assays to
aid in predicting at an early stage of development not only the
intrinsic activity of compounds, but also their potential side
effects such as impact on important non-target species including
beneficial insects and vertebrates.

In some cases, such automated biochemical methods may help in
the discovery process since many more samples can be run through
assays for AChE inhibitors, receptor binding, Hill reaction
inhibitors, and metabolism inhibitors as potential synergists than
can be put through a normal biological screen in the same time
period. As long as such biochemical screens are relevant for the
biological activity desired (43), and are not meant to replace whole
animal screening, but simply to augment it and provide information on
potentially interesting intrinsic activity for further evaluation,
their utility may be considerable. For example, such automated
screening with AChE as the target resulted in the discovery of a
novel series of inhibitors (51), although unfortunately with
selective activity against mammalian forms of the enzyme.

Discovery of the mode and site of action of existing compounds. This
is a second way in which studies on pest biochemistry and physiology
can aid in the discovery of new and safer compounds. The discovery of
the site and mechanism of action of a group of pesticides inevitably
stimulates new thinking, particularly if the site is a novel one.

A recent example from the area of herbicides illustrates this possibility very well. Two recent new groups of herbicides with very high commercial potential are the sulfonylureas and imidazolinones. It was found that, though quite different structurally, these compounds inhibit the same enzyme in plants, acetohydroxyacid synthase (AHAS), which is essential for the synthesis of the branched chain amino acids valine, leucine and isoleucine. The mechanism of inhibition is still under study, but probably both groups of compounds act on a regulatory site rather than on the active site of the enzyme (52). Valine itself is known to exert feedback inhibition on the activity of some forms of AHAS. Using this information on the potential site and mode of action, Huppatz and Casida (53, Figure 3) noted that the imidazolinones contain a 2-methylvaline substructure and postulated that other valine derivatives might therefore also interact at the regulatory site to inhibit AHAS activity. A series of valine analogs were synthesized, and the most active one, N-phthaloyl-L-valine anilide, was found to inhibit AHAS and plant growth in the micromolar range.

It must be stressed that the contributions of pesticide mode of action studies are not limited to their undoubted practical significance in pesticide discovery. The definition of modes of action in pests generally has considerable significance in the safety evaluations of a pesticide. Further, the elucidation of a new mechanism of action may produce remarkable benefits for basic biological research. It frequently stimulates research in that field, often generates new insights into essential biological processes, and provides indispensible tools for their study. It would be safe to say that many of the most significant advances in modern biology depended on the use of natural and man-made poisons as probes.

An even higher degree of sophistication in pesticide discovery is now within sight. The increasingly detailed understanding of enzyme mechanisms sometimes allows the design of inhibitors that interact very strongly or irreversibly as well as specifically with them e.g. transition state analog inhibitors and suicide substrates (48). Remarkable advances in biochemical and molecular genetic techniques and X-ray crystallography allow the amino acid sequences and prediction of 3-dimensional structures of important target sites to be generated with increasing facility. The computer-assisted definition of the structure and chemistry of binding domains of potential target receptors and enzymes and of their interactions with ligands is already being practiced (43,54,55). This is discussed by Dr. Vorpagel in another chapter in this volume.

The best developed example of this process with pesticides currently lies in the herbicide area. Triazines, ureas, and many other herbicides inhibit photosynthesis by competitively displacing plastoquinones from their binding site in photosystem II, thus diverting electron flow from its normal pathway and resulting in the generation of reactive intermediates that are cytotoxic (19,52). Recently the position of this binding site has been localized to a 32 kDa polypeptide and the molecular architecture of the site has been elucidated using such techniques as photoaffinity labelling, gene sequencing of herbicide-resistant mutants, the development of a plausible model of the amino acid sequence and its folding within the membrane, and comparison with the structure of the related region in the bacterial photosystem determined by X-ray crystallography

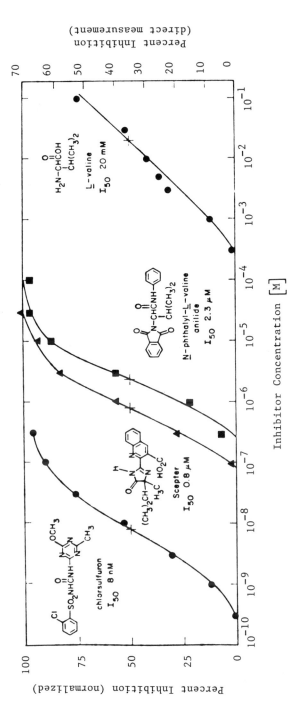

Figure 3. Relative potencies of N-phthalyl-L-valine anilide, L-valine, the imidazolinone herbicide, Scepter, and the sulfonylurea herbicide, chlorsulfuron, as inhibitors of acetohydroxyacid synthase from Zea mays. (Reproduced with permission from Ref. 53. Copyright 1985 Verlag der Zeitschrift fur Naturforschung.)

(52,56,57; Figure 4). This information is now being used in industry
to design novel inhibitors for this site as potential herbicides.
   Returning to the topic of ACh receptors, preliminary three-
dimensional models are available for this receptor from Torpedo
electroplax (58,59). Figure 5 shows the possible structure of the
ACh binding site and the residues likely to be involved in ACh
binding. Considerable progress can be expected in developing the
structure of the comparable receptor from the locust CNS (60).
Notable differences exist in the subunit structure of these two
receptors, although they may not extend to all types of mammalian
ACh receptors. Information of this type may eventually allow the
design of agonists and antagonists having a high degree of
specificity for insect receptors. In addition to revealing the
topography of the ACh binding site, it is possible that such studies
will reveal new binding areas at which ligands could either trigger
or inhibit the normal receptor-activated opening of ion channel. The
ion channel itself offers such an opportunity since several
pharmacologically active agents have been shown to interact with this
pore (58) including some insecticides (61).
   In the area of fungicides, the binding site for benzimidazole
fungicides on fungal beta-tubulin in both sensitive and resistant
fungi is now beginning to be understood at the molecular level (28).
Additionally, a group of enzymes with great significance for
pesticide action and resistance is the cytochrome P-450 based family
of monooxygenases. These act both as a major metabolic force for
pesticides, and as a potential target site e.g. they are the
established site of action for many fungicidal ergosterol
biosynthesis inhibitors and a potential site for inhibitors of
juvenile hormone biosynthesis in insects (19). It is therefore
significant that the structure of one form of P-450 has recently been
revealed by X-ray crystallography (62). Further developments along
these lines could eventually open the way for the computer-assisted
design of several types of pesticides, growth regulators and
pesticide synergists. Applications of this approach to aid in
developing inhibitors of sterol biosynthesis in fungi (63) and
gibberellin biosynthesis in plants (64) have been described.
   The information regarding these and other binding sites should
be invaluable for the design of improved and new types of ligands.
At present the discovery and optimization of effectors is very much a
hit and miss phenomenon, akin to the old game of pinning the tail on
a donkey while blindfolded. To have the 3-dimensional structure and
reaction mechanism of a critical site fully elucidated is to remove
the blindfold. To appreciate the potential of this site-directed
molecular design one has only to remember that, by adding to an
existing molecule a strategically located functional group which
forms an additional hydrogen bond with the receptor surface, one can
increase its affinity by two or three orders of magnitude (65).
Unfortunately, knowledge of most potential target sites at this
sophisticated level in most target species is very limited or
completely lacking and a great deal of effort will be needed to
develop it to a usable level.

New Concepts in Pest Control - The Discovery of Pest Vulnerability.
Further back in the chain of discovery, but of extreme importance for
future developments, studies of pest biology, physiology and

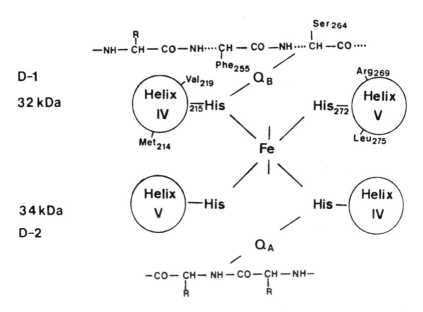

Figure 4. Proposed plastoquinine (QB) and herbicide binding site on the 32 kDalton D-1 polypeptide of photosystem II. The quinone is bound through an iron-complexed histidine residue (his 215) and hydrogen bonding to ser 264. Further interactions occur with arg 269 and phe 255 lying above and below the binding site. Amino acid substitutions in herbicide-tolerant mutants have been identified at the residues numbered 219, 255, 264 and 275. Reproduced with permission from Ref. 57. Copyright 1986 Verlag der Zeitschrift fur Naturforschung.

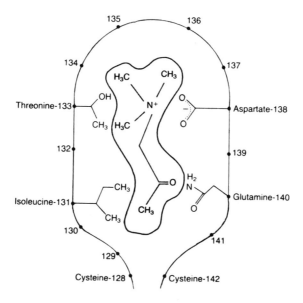

Figure 5. Model of the proposed acetylcholine recognition site on the alpha subunit of the Torpedo ACh receptor in the region between Cys 128 and Cys 142. An ACh molecule is shown in relation to the four residues postulated to interact in its binding. Reprinted with permission from Ref. 59. Copyright 1985 Elsevier Science Publishers.

biochemistry lead to new concepts in pest control by defining basic differences between target and non-target species. As examples, the discovery of the control of insect development by juvenile hormones, and of mating and aggregation behaviors by insect pheromones, were eagerly and rapidly exploited for new and very safe approaches to insect control. These areas continue to attract research aimed at novel applications of this basic knowledge. Even restricting the discussion to insect pests, many other significant differences in their biochemistry and physiology compared to vertebrates are already known. These represent potential areas for the development of safer pesticides, some of which are currently under study with this end in view e.g. peculiarities of sterol metabolism in insects, their reliance on octopamine and specific neuropeptides as neuroeffectors, and their cuticular biochemistry and waterproofing (45,46,48,66). Further, selectivity and mammalian safety can be achieved just as readily by the development of compounds which are degraded at differential rates in friend and foe as by attacking biochemical targets peculiar to the pest, as illustrated by the pyrethroid insecticides. Lack of ability to rapidly degrade a toxicant is clearly a serious vulnerability factor. There is a great need for studies on the basic physiology and biochemistry of all types of pests to continue and be expanded in order to better understand existing sites of vulnerability and to discover new ones which can lead to novel concepts for control.

To provide but a single example from many, the defense responses of insects against invading pathogens represents just such a promising topic of current research. Results in this area have recently been reviewed by Dunn (67). In lepidopterous larvae, these defenses are multiple and include both humoral and cellular elements. Invading organisms are subject to phagocytosis and encapsulation in cellular networks derived from hemocytes. Over a period of 24 to 48 hours, these initial defenses are augmented by the synthesis and release into the blood of lysozyme and lysine- and arginine-rich polypeptide antibiotics (bactericidins) from the fat body. It is premature to suggest that such information currently provides obvious opportunities for exploitation in insect control, but the potential is clear. If means could be found to turn off or obviate some or all of these defenses, the insect would be more likely to succumb to natural infections. This could be particularly useful if applied in combination with an infectious agent such as a bacterium or virus. Such a concept is more than a pipe-dream since it has been shown that a virus associated with the eggs of parasitic wasps prevents the encapsulation response of the host which would otherwise tend to occlude the egg (68). An unidentified component from the venom of other parasitic wasps has also shown immunosuppressant properties in insects (69) and two immunosuppressants active on giant silkworm moths have been detected in preparations of Bacillus thuringiensis (70). Further, Dunn and his coworkers have found that quite small peptidoglycan fragments of the bacterial cell wall are recognized as "foreign" in Manduca sexta and will trigger the production of humoral defenses. Can we find low molecular weight analogs to block these "receptor" sites and thus inhibit this aspect of immunity? Finally, the defensive antibiotics are rich in arginine. Canavanine, an analog of arginine which occurs in some legume seeds, can be toxic by replacing arginine in essential proteins and peptides (71).

$$NH$$
$$H_2N-\overset{\parallel}{C}-NHOCH_2CH_2CH(NH_2)COOH$$

Canavanine

Manduca larvae treated with canavanine produce antibiotic peptides in
which arginine is replaced by canavanine. Their intrinsic biological
activity is greatly diminished and these larvae may be unusually
susceptible to infection by bacteria (P. E. Dunn, personal
communication).

It is only realistic to admit that the chances of eventually
being able to exploit any such example of basic research are highly
uncertain and require a long lead time for development. Even the
discovery of mechanisms to disrupt essential biochemical processes
peculiar to pest groups does not offer a panacea for several reasons.

1. Though we may avoid acute toxicity to most non-target species,
there are no guarantees that we will avoid chronic toxicity: the two
are not correlated. In fact there may be a negative correlation since
current protocols for safety evaluation employ the maximum tolerated
dose in long term studies with vertebrates. Hence compounds which can
be given at high doses because of their low acute toxicity are at
a disadvantage in lifetime chronic tests.

2. Though the site of action may be peculiar to the pest, there
may be parallel processes in non-target species that also are
affected.

3. Even if after enormous effort and investment a new approach
clears all these hurdles and can be used in practice, we shall find
no victory over pests to be permanent. The loss of vulnerability
through the development of resistance will always be a threatening
possibility. This is why it is so crucial that much greater effort be
devoted to understanding the population, genetic, and biochemical
basis of the development of resistance to control measures in
general.

However, on the basis of past experience, it is reasonable to
conclude that with faith, patience, and a sufficient investment, at
least some such basic studies of pest biochemistry will pay off in
novel and safer pest control strategies.

The Clouded Future

Since the Second World War the US has invested large amounts of money
on research into basic biological mechanisms and their control,
largely spurred by medical goals, but also, to a lesser extent, by
the motivation to understand potential pests and their essential
characteristics. This investment by government and industry has
proved remarkably fruitful. There is reason to hope that we are
now on the threshold of an era where we will be able to design
compounds that are both highly potent and highly specific in their
toxic effects, and to prevent resistance from negating them. Although
much remains to be done to achieve these ends there is cause for
optimism that we can achieve them. However, such advances, starting
from basic studies, are increasingly costly and demand a team effort

and cooperation between industry and academia. This comes at a time when the agrochemical industry is experiencing lowered profits and a decreased ability to invest in such basic and expensive research. A third essential partner in this process is government, particularly, because of its agricultural mandate, the USDA. However, governmental research support is also experiencing financial stringency and uncertainty.

The USDA, even in the past, has not, from the perspective of this article, invested its limited resources optimally. When it became clear in the 1960's that society correctly regarded many current pesticides as too flawed for continued use, two logical approaches to the development of safer pest control strategies could be envisioned: (1) to develop alternatives to chemical pesticides, and (2) to support basic research aimed towards the discovery of new and safer chemical compounds to replace those that were unsatisfactory. The USDA chose to emphasize the former approach heavily at the expense of the latter. Though politically popular, this may have been short-sighted, and has not radically changed our approach to pest control which continues to depend very heavily on pesticides. Meanwhile funding from this potentially primary source to support research into the better understanding of the biochemical sites and modes of action of pesticides, to aid in the discovery of new methods to develop improved and safer compounds, and to spur efforts to understand and alleviate resistance to pesticides has been negligible in comparison to the needs and opportunities. This viewpoint has been independently expressed by others also (48). It is to be hoped that the emphasis on strategies for developing safer pest control technologies can soon be brought into a more rational balance.

When one considers the immense sums being invested in weapons and medical research, or elsewhere in agriculture, the amount needed to support such work at a realistic level is miniscule. Just 0.01% of the amount expected to be spent to subsidize agricultural production in the USA in 1986 would provide $3 million as a firm foundation for pesticide-related basic research. This same amount represents 2% of what is spent to maintain military bands in the USA. Currently, there is a minimum of encouragement or opportunity in the USA to work on these topics, and a large proportion of the research is done in other nations. At the recent IUPAC International Congress of Pesticide Chemistry, which attracted a worldwide attendance and was held close to the USA in Canada, 78% of the papers on fungicide modes of action were given by speakers with non-US affiliations, as were 89% of the papers on herbicide modes of action, and 82% of those on fungicide and insecticide resistance. While it is encouraging that such work is thought worthy of support in many nations around the world, the clear indication is that, despite its huge agricultural industry with a continuing dependence on pesticides, the USA is lagging seriously in such necessary research efforts.

The combination of financial stresses that currently threaten all sources of funding for agricultural research is in danger of leaving us with fruit ripe for the picking as a result of our prior investments in biological research, but with few harvesters in the orchard. If so, others may then profit from our lack of vision.

Acknowledgments

My thanks go to Drs. P. E. Dunn and J. V. Osmun for critically
reading this manuscript, and to Dr. M. E. Schroeder for providing
unpublished information regarding the discovery of the nitromethylene
heterocyclic insecticides. Journal Paper No. 10994, Agricultural
Experiment Station, Purdue University, W. Lafayette, IN 47907.

Literature Cited

1. Hayes, W. J., Jr.; Vaughan, W. K. Toxicol. Appl. Pharmacol.
   1977, 42, 235-52.
2. Copplestone, J. F. In "Pesticide Management and Insecticide
   Resistance"; Watson, D. L.; Brown, A. W. A., Eds.; Academic:
   New York, 1977; pp. 147-55.
3. Jeyaratnam, J.; de Alwis Seneviratne, R. S.; Copplestone, J. F.
   Bull. World Health Org. 1982, 60, 615-9.
4. Davies, J. E.; Doon, R. In "Silent Spring Revisited"; Marco,
   G. J.; Hollingworth, R. M.; Durham, W. W., Eds.; American
   Chemical Society: Washington, D.C., 1986, In Press.
5. Davies, J. E.; Freed, V. H.; Whittemore, F. W. "An Agromedical
   Approach to Pesticide Management"; Univ. Miami Press, 1982; pp.
   8-9.
6. Pimentel, D. In "Chemistry and World Food Supplies: The New
   Frontiers"; Shemilt, L. W., Ed.; Pergamon: Oxford, 1983, pp.
   185-201.
7. Barthel, E. J. Toxicol. Environ. Health 1981, 8, 1027-40.
8. Hoar, S. K.; Blair, A.; Holmes, F. F.; Boysen, C. D.; Robel, R.
   J.; Hoover, R.; Fraumeni, J. F., Jr. J. Am. Med. Assoc. 1986,
   256, 1141-7.
9. Hall, R. J. In "Silent Spring Revisited"; Marco, G. J.;
   Hollingworth, R. M.; Durham, W. W., Eds.; American Chemical
   Society: Washington, D.C., In Press.
10. Knipling, E. F. "The Basic Principles of Insect
    Population Suppression and Management"; Agric. Handbook No.
    512; U.S. Dept. Agric.: Washington, D.C., 1979, pp. 425-7.
11. Campion, D. G. In "Techniques in Pheromone Research"; Hummel,
    H. E.; Miller, T. A., Eds.; Springer-Verlag: New York, 1984,
    pp. 405-49.
12. Hollingworth, R. M.; Lund, A. E. In "Insecticide Mode of
    Action"; Coats, J. R., Ed. Academic: New York, 1982, pp.
    189-227.
13. Hollingworth, R. M.; Lund, A. E. In "Pesticide Chemistry:
    Human Welfare and the Environment"; Miyamoto, J.; Kearney,
    P. C., Eds.; Pergamon: Oxford; 1983, Vol.3, pp. 15-24.
14. Nathanson, J. A. Science 1984, 226, 184-7.
15. Griffiths, D. C.; Pickett, J. A. Entomol. Exp. Appl. 1980,
    27, 199-201.
16. Hess, F. D.; Bayer, D. E.; Falk, R. H. Weed Sci. 1981, 29,
    224-9.
17. Schouest, L. P., Jr.; Umetsu, N.; Miller, T. A. J. Econ.
    Entomol. 198, 76, 973-82.
18. Pallos, F. M.; Casida, J. E. "Chemistry and Action of Herbicide
    Antidotes"; Academic: New York, 1978.

19. Corbett, J. R.; Wright, K.; Baillie, A. C. "The Biochemical Mode of Action of Pesticides"; Academic: New York, 2nd Edn., 1984.
20. Menn, J. J. J. Agric. Food Chem. 1980, 28, 2-8.
21. Brattsten, L. B.; Holyoke, C. W., Jr.; Leeper, J. R.; Raffa, K. F. Science 1986, 231, 1255-60.
22. Shaaltiel, Y.; Gressel, J. In "Pesticide Science and Biotechnology"; Greenhalgh, R.; Roberts, T. R., Eds.; Blackwell: London, In Press.
23. Georghiou, G. P.; Saito, T. "Pest Resistance to Pesticides"; Plenum: New York, 1983.
24. Dover, M.; Croft, B. "Getting Tough: Public Policy and the Management of Pesticide Resistance"; World Resources Institute: Washington, D.C., 1984.
25. "Pesticide Resistance: Strategies and Tactics for Management", National Research Council, National Academy Press, Washington, D.C., 1986.
26. Sawicki, R. M. Phil. Trans. R. Soc. Lond. B 1981, 295, 143-51.
27. Oppenoorth, F. J. In "Comprehensive Insect Physiology, Biochemistry and Pharmacology"; Kerkut, G. A.; Gilbert, L. I., Eds.; Pergamon: Oxford, 1985; Vol. 12, pp. 731-73.
28. Davidse, L. C. Ann. Rev. Phytopathol., 1986, 24, 43-65.
29. Miyata, T. In "Pest Resistance to Pesticides"; Georghiou, G. P.; Saito, T., Eds.; Plenum: New York, 1983; pp. 99-116.
30. Pasteur, N.; Georghiou, G. P. Mosquito News 1981, 41, 181-3.
31. Ali, A.; Souza Machado, V. Weed Res. 1981, 21, 191-7.
32. Devonshire, A. L.; Moores, G. D.; Ffrench-Constant, R. H. Bull. Entomol. Res. 1986, 76, 97-107.
33. Hemingway, J.; Rubio, Y.; Bobrowicz, K. E. Pestic. Biochem. Physiol. 1986, 25, 327-335.
34. Georghiou, G. P. In "Pest Resistance to Pesticides"; Georghiou, G. P.; Saito, T., Eds.; Plenum: New York, 1983; pp. 769-92.
35. Ozaki, K. In "Pest Resistance to Pesticides"; Georghiou, G. P.; Saito, T., Eds.; Plenum: New York, 1983; pp. 595-614.
36. de Waard, M. A.; van Nistelrooy, J. G. M. Pestic. Sci. 1983, 15, 56-62.
37. Yamamoto, I.; Takahashi, Y.; Kyomura, N. In "Pest Resistance to Pesticides"; Georghiou, G. P.; Saito, T., Eds.; Plenum: New York, 1983; pp. 579-94.
38. Yamamoto, I. Abst. 6th IUPAC Int. Cong. Pestic. Chem., Ottawa, Canada, 1986. Abst. 3E-20.
39. Elliott, M.; Farnham, A. W.; Janes, N. F.; Johnson, D. M.; Pulman, D. A.; Sawicki, R. M. Agric. Biol. Chem. 1986, 50, 1347-9.
40. Nakamura, S.; Kato, T.; Noguchi, H.; Takahashi, J.; Kamoshita, K. In "Pesticide Science and Biotechnology"; Greenhalgh, R.; Roberts, T. R., Eds.; Blackwell: London, In Press.
41. Uesugi, Y. In "Pest Resistance to Pesticides"; Georghiou, G. P.; Saito, T., Eds.; Plenum: New York, 1983; pp. 481-504.
42. Braunholtz, J. T. Phil. Trans. R. Soc. Lond. B, 1981, 295, 19-34.
43. Geissbuehler, H.; Mueller, U.; Pachlatko, J. P.; Waespe, H. R. In "Chemistry and World Food Supplies: The New Frontier"; Shemilt, L. W., Ed.; Pergamon: Oxford, 1983; pp. 643-56.

44. Storck, W. J. Chem. Eng. News 1984, 62 (April 9), 35-57.
45. Magee, P. S.; Kohn, G. K.; Menn, J. J. "Pesticide Synthesis through Rational Approaches"; ACS Symposium Series No. 255; American Chemical Society: Washington, D.C., 1984.
46. von Keyserlingk, H. C.; Jaeger, A.; von Szczepanski, C. "Approaches to New Leads for Insecticides"; Springer: Berlin, 1985.
47. Schwinn, F.; Geissbuehler, H. Crop Protect. 1986, 5, 33-40.
48. Hammock, B. D.; Soderlund, D. M. In "Pesticide Resistance: Strategies and Tactics for Management"; National Academy: Washington, D.C., 1986; pp. 111-29.
49. Schroeder, M. E.; Flattum, R. F. Pestic. Biochem. Physiol. 1984, 22, 148-60.
50. Soloway, S. B.; Henry, A. C.; Kollmeyer, W. D.; Padgett, W. M.; Powell, J. E.; Roman, S. A.; Tieman, C. H.; Corey, R. A.; Horne, C. A. In "Advances in Pesticide Science; Geissbuehler, H., Ed.; Pergamon: Oxford; 1979, Pt. 2, pp. 206-17.
51. Voss, G.; Neumann, R. Experientia 1979, 35, 583-4.
52. Fedtke, C.; Trebst, A. In "Pesticide Science and Biotechnology"; Greenhalgh, R.; Roberts, T. R., Eds.; Blackwell: London, In Press.
53. Huppatz, J. L.; Casida, J. E. Z. Naturforsch. 1985, 40C, 652-6.
54. Gund, P.; Andose, J. D.; Rhodes, J. B.; Smith, G. M. Science 1980, 208, 1425-31.
55. Richards, W. G. Endeavour 1984, 8, 172-8.
56. Deisenhofer, J.; Epp, O.; Miki, K.; Huber, R.; Michel, H. Nature 1985, 318, 618-24.
57. Trebst, A. Z. Naturforsch. 1986, 41C, 240-5.
58. Changeux, J.-P.; Devillers-Thiery, A.; Chemouilli, P. Science 1984, 225, 1335-45.
59. White, M. M. Trends in NeuroSci. 1985, 8, 364-8.
60. Breer, H.; Kleene, R.; Hinz, G. J. Neurosci. 1985, 5, 3386-92.
61. Eldefrawi, M. E.; Sherby, S. M.; Abalis, I. M.; Eldefrawi, A. T. Neurotoxicol. 1985, 6, 47-62.
62. Poulos, T. L. In "Cytochrome P-450"; Ortiz de Montellano, P. R., Ed.; Plenum: New York, 1986; pp. 505-23.
63. Marchington, A. F. Proc. 10th Internat. Congr. Plant Protect. 1983, 1, 201-8.
64. Ebert, E.; Huxley, P.; Mueller, U. Abst. 6th IUPAC Internat. Congr. Pestic. Chem. Ottawa, Ontario, Abst. #1C-15.
65. Andrews, P. Trends Pharm. Sci. 1986, 7, 148-51.
66. Hollingworth, R. M. In "Pesticide Selectivity"; Street, J. C., Ed.; Dekker: New York; 1975, pp. 67-111.
67. Dunn, P. E. Ann. Rev. Entomol., 1986, 31, 321-39.
68. Stoltz, D. B. J. Insect Physiol., 1986, 32, 347-50.
69. Kitano, H. J. Insect Physiol., 1986, 32, 369-75.
70. Edlund, T.; Siden, I.; Boman, H. G. Infect. Immun. 1976, 14, 934-41.
71. Rosenthal, G. A. Q. Rev. Biol. 1977, 52, 155-78.

RECEIVED December 5, 1986

Chapter 7

Pests as Part of the Ecosystem

L. V. Madden

Department of Plant Pathology, The Ohio State University, Wooster, OH 44691

Diseases, insects, and weeds are important constrains
to crop production; their combined effect has been
"questimated" at 25-45%, on a world-wide basis.
Pesticides comprise one of the major means of
pests. In the U.S., insecticides, herbicides, or
fungicides are used on more than 90 million hectares
of crop land. Percentages of crop area treated with
pesticides range from <1 to >90% for crops ranging
from pasture to apples. The amount of chemicals used
and the seriousness of plant pests mandate that
pesticide usage be based on sound biological data and
principles. Pest control decisions ideally are based
on: 1) precise estimates of pest density and
aggregation; 2) interaction of pests with crop plants
and the environmental and biotic environment; 3)
quantification and modeling of pest population dynamics;
and 4) effects of pests on crop yields. Optimal timing,
amount, and selection of pesticides (e.g., type of
activity), as well as selection of alternate controls,
can be determined based on these four considerations.
Pesticide risks to humans as well as the environment
can be reduced through an understanding of the pest
as part of the ecosystem.

Pests are inescapable parts of an ecosystem. No crop can be grown
anywhere in the world without concern about damage due to at least
one pest. For this paper, the term pest is used in a gross sense to
include all organisms that are detrimental to agricultural
production, weeds, insects and other arthropods (e.g., mites), and
pathogens. Pathogens include fungi, viruses, nematodes, bacteria,
and other prokaryotes such as spiroplasma and mycoplasma like
organisms (MLO's) that cause plant diseases. Pathogens of crop
plantsare dominated by fungi, although the other groups are
extremely important. There are other pest types such as mammals
(e.g., deer, rodents) and birds, but these are somewhat less
widespread and will not be specifically discussed any further.
    The impact of pests is immense but quantitative data is
notoriously poor (1). In an attempt to understand losses caused by

0097-6156/87/0336-0077$06.00/0
© 1987 American Chemical Society

plant diseases, Zadoks and Schein (2) classified losses as being
direct or indirect; their classification system is pertinent to all
pests and not just pathogens. Direct losses are reductions in
quantity or quality of produce, as well as reduction of yielding
capacity. Direct losses include: losses incurred pre- and
postharvest and costs of control (primary direct losses);
contamination of sowing and plant material; soil infestations, and
other factors that reduce yield in future seasons (secondary direct
losses). Indirect losses encompass economic and social effects of
pests, including economic impacts on farmers, communities,
consumers, and the environment.

     In general, agricultural scientists cannot give accurate
estimates of the above losses. This is due partly to the lack of
detailed field experiments relating pest levels to loss, and partly
to the lack of extensive survey data on the level of pests in
grower's fields. The published estimates of losses are better
called "questimates" and should only represent general ranking of
pest effects. On a world-wide basis, primary losses due to pests
are in the range of 35% (3), fairly evenly attributable to insects,
diseases and weeds (Table I).

Table I.   Some Published Percentage Losses Due to Pests
            (reported in McEwen (3))

	Insects	Diseases	Weeds	Total
North America	9	11	8	28
World	11.6	12.6	10	34.2
Wheat	5	9	10	24
Potato	5	22	4	31
Rice	27	9	11	47

Other published values may be found in the literature, but the
values presented herein likely are just as accurate. In North
America, losses are somewhat less than for all continents. These
numbers for continents, however, mask many interesting results. For
instance, diseases and weeds cause more losses in wheat than do
insects; the opposite is true for rice (Table I). Obviously, these
losses will vary with location, year, and cropping practices.

     Due to impact of pests, considerable expenses are incurred in
controlling insects, weeds, and pathogens. One such control
practice is the use of pesticides. In the U.S., over 90 million
hectares are treated with herbicides, as an example (4),
considerably less insecticides and fungicides are used (Table II).
In 1982, herbicides accounted for more than three-quarters of all
pesticide sales.

Table II. Pesticide Use on Major Food Crops in the U.S. for
1982 based on Schaub (4)

Pesticide	Amount (million kg a.i.)	Hectares (millions)
Herbicides	190.7	90.0
Insecticides	24.8	21.9
Fungicides	2.4	1.5

Hectares covered and the amount used vary tremendously with
crops and locations. For instance, in 1976 nearly 50% of the
insecticide use on major field crops was on cotton (4). The
greatest fungicide use is on fruit and vegetable crops. However,
this use varies with geographic location. For example, although
~50% of potatoes in the U.S. are sprayed with fungicides, but close
to 100% are sprayed in the eastern U.S. (5).
    The above figures should indicate that pests have a serious
impact on crop production and that pesticides comprise a major means
of control. To minimize risks due to pesticide applications, a
clear understanding of the biology, ecology, and population dynamics
of pests is imperative. It is no longer economical or
environmentally desirable to apply excessive and poorly timed
chemicals. Pesticide application strategies should be based on: 1)
precise estimates of pest density and aggregation; 2) interaction of
pests with crop plants and the physical and biotic environment; 3)
quantification and modeling of pest population dynamics; and 4)
effects of pests on crop yield.

Pest Density and Aggregation

A trivial principle of biology is that organisms are not equal in
number at all locations. This infers that plant pests are not
equally important in all geographic regions. For instance, one of
the most serious constraints to potato production in the eastern
U.S. is the disease, late blight, caused by the fungus Phytophthora
infestans. Without regular fungicide applications, potatoes could
not be grown. However, in many western U.S. states the disease is
not of major concern, mainly because of much drier weather
conditions. Another example is the weed johnsongrass (Sorghum
halepense), a major pest in the southern U.S., which is not found in
high numbers in the northern U.S. This large-scale geographic
variation in pest numbers and impact results in differing control
practices.
    At a much smaller scale, pest density can vary tremendously
within a given region. This is particularly true with soil-borne
fungi and arthropods. Two adjacent fields may have densities that
vary by several orders of magnitude. With pests that move in the
air, there will be less aggregation than this, but differences can
still be large. Unfortunately, it is still not a wide-spread
practice to assess pest densities in fields in order to properly
determine the need to use a pesticide. There are obvious
exceptions, however, which fall under the concept of Integrated Pest

Management (IPM). Initiated in the 1950's to better control insects of cotton, IPM blossomed in the 1970's with a wealth of federal government (USDA, NSF, and EPA) backing (6-7). IPM encompasses a holistic, multidisciplinary management system that integrates control methods on the basis of ecological and economic principles for pests that coexist in an agroecosystem. It involves much more than assessing pest density, but these assessments, nevertheless, are critical. Although pests, especially insects, of cotton and tobacco were originally studies in pilot projects, many crops in many states were eventually included.

An implicit assumption of IPM is that pesticides should only be used when necessary. Absence, or anticipated absence, of a given pest is a situation in which a pesticide application is not necessary. Even when a pest is present, control decisions can be made on the anticipated pest increase and the relationship between pest numbers and yield (see sections below). In practice, trained scouts can sample a given field by counting, measuring, or assessing pest density at selected locations. Growers themselves can also make the assessment.

The degree to which IPM is practiced and scouts are used varies with the crop and geographic area. Scouts are heavily used in some states (e.g., California) and rarely used in others (e.g., Ohio). Although usually lower expenditures for pesticides are required with IPM, sometimes greater expenditures are necessary in favorable years (4). For example, apple growers participating in an IPM program in North Carolina used less insecticides but more fungicides, resulting in an increase in total pesticide expenditures (8). However, increased fruit quality produced a net revenue increase for IPM growers compared to non-IPM growers.

One generally wants to determine the mean pest density per field with a given degree of precision. To do this, one must take a number of samples that is dependent on the aggregation of the pest. At any given time, a pest might have a uniform, random or clustered pattern in a field (9) (Fig. 1). Uniform patterns are not expected to occur, and even random patterns are not common. Although randomness is an absolute, clustering is a matter of degree--from low to high clustering. For a given level of precision, the lowest number of required samples is with a random pattern. Sample size increases as clustering increases.

There are many ways to assess or characterize the degree of clustering (9-10). For our purposes, the simplest measurement will be presented. If one is counting number of individual insects, infected plants, or lesions, and some statistical assumptions are met, the variance (v) will equal the mean (m) if there is a random pattern. With clustering, $v>m$, or the variance-to-mean (VTM) ratio exceeds one. The VTM is a useful yet simple index of aggregation; if VTM is known, one can sample accordingly. Obviously, the exact degree of aggregation will not be known until after sampling, but based on prior studies, one can assume a "worst-case" scenario and collect the necessary number of samples. Formulae for sample sizes are available (9).

Interaction of Pests with the Environment

Insects, pathogens, and weeds respond to their physical and biotic environment in predictable ways. For instance, growth of many fungal pathogens varies with temperature in a well-established manner. Growth starts low, increases to a maximum at the optimal temperature and then declines to zero (Fig. 2). In fact, most plant pests will respond to temperature in a similar manner. Fungal pathogens often require free moisture for infection to occur; infection increases with increased time of wetness (Fig. 2). The duration of free moisture is dependent on temperature (Fig. 2), as well as other physical factors.

Development of insects and pathogens is highly dependent on the plant host cultivar. Both physical and physiological host factors influence or limit the life cycle of insects and the disease cycle of pathogens. In fact, the first line of defense against many pathogens is host resistance. Although chemicals are more often used to control insects, host resistance to insects can be very effective (11).

Figure 2 depicts just some of the known response-stimulus relationships for plant pests. Collection of the fundamental data and description of the responses with mathematical models can lead to a better understanding of pests in the field. If one knew the current environmental conditions, one could predict whether or not the potato leafhopper would increase during the next week. Final decisions would also be based on the known population dynamics of the pests of interest (see below), as well as the known interaction between the plant host and pest.

Another component of the biotic environment is the collection of organisms that interact with the pests. Some of these interactions, including competition, parasitism, and predation, are exploited by man to achieve biological control. The range and numbers of interactions are immense. Researchers are accustomed to conducting experiments with perhaps two or three interactions. But when considering pests as part of the ecosystem, there are thousands of interactions (or relationships) among organisms within and between crops, as well as with crop cultivars, cultural conditions, and the physical environment. It is revealing to note that most pesticide use is aimed at curtailing interactions by the high specificity of the utilized chemicals. Van Enden (12) discusses some interesting interactions of pesticides with other factors, including biological control agents. An area in which interactions are critical for crop management is minimum or no-tillage systems. Here, interactions in the soil are relied on to result in reduced pest levels. The long-term consequences of the use of herbicides with minimum tillage cropping systems is still not known. Perhaps pest interactions should be exploited rather than eliminated with pesticides to benefit growers. Obviously much more work needs to be done in this area.

Population Dynamics

Numbers of pests are seldom static for long periods of time. Usually, population growth can be described precisely with one or more mathematical models. Such models permit the prediction of

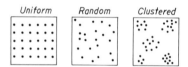

Fig. 1. Example patterns of pests in field plots.

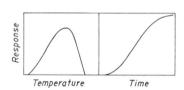

Fig. 2. Typical relationships between pests and physical factors.

future population levels based on past population growth and current
environmental and biotic conditions. Knowledge of pest population
dynamics is paramount for correctly using pesticides. Knowing that
a certain pest will or will not reach a level at which losses occur
can completely determine whether or not a pesticide is applied.
Pests that increase slowly can be efficiently controlled at planting
time, whereas pests increasing at a high rate need to be controlled
throughout most of the growing season (2).
    There is a wealth of literature on pest population dynamics
(e.g., 13-14). Only some rudimentary concepts can be given here.
The growth of a population is a dynamic process that needs to be
described by its rate of change with time (dY/dt):

$$dY/dt = rY(K-Y)/K \qquad\qquad (1)$$

in which: Y is the number of pests (e.g., insect adults, fungal
lesions, or even infected plants); t is time, r is a rate parameter;
K is the maximum population size, i.e., the carrying capacity; and
dY/dt represents the absolute rate of increase. Integrating this
equation results in a sigmoid-shaped curve which analytically can be
written as:

$$Y = K/(1+\exp(-(\ln(Yo/(K-Yo))+rt)) \qquad\qquad (2)$$

in which Yo represents initial population size. Equations 1 and 2
are very simple models for population growth, but, nevertheless, are
often used, both for prediction and interpretation. For instance,
it can be shown that reducing Yo with some control measure is not
efficient if r is high (2).
    Equation 1 can be generalized considerably to make it more
realistic. The rate parameter r is influenced by generation time,
reproductive and disperal capacity, and how the biotic and abiotic
environment influences these components. One can refine eq. 1 to
account for all of these aspects by dividing Y into reproductive and
nonreproductive parts (e.g., latent and infectious lesions), and
also by making r a variable that is a function of the environment
K can also be made a variable that can vary with host changes and
other factors.

Effects of Pests on Crop Yields

Obviously, pests would not be important if they did not reduce crop
yields. A great deal of recent work has been conducted to relate
pest density at various times to yield loss (e.g., 15-17). Pests
can be classified into various categories, including: stand
reducers (e.g., damping-off fungi), photosynthetic rate reducers
(viruses), leave senescence accelerators (pathogens), light stealers
(weeds), assimilate sappers (nematodes, sucking insects), tissue
consumers (chewing insects, fungi) and turgor reducers (root feeding
insects and pathogens) (15). Any given pest can act in one or more
of the above categories.
    The combined effects of plant pests will result in measurable
direct primary loss if a threshold pest density is surpassed. Below
this level (injury or disease damage level), crops theoretically are
capable of "compensating" for the injurious effects of the pests.

Such thresholds, however, are difficult to determine experimentally. For instance, at low pest density (where the threshold is likely to be) there is great within-field variation and pest aggregation. Taken together with the natural variation in yield among healthy plants and the fact that other factors such as weather and nutrient availability influence results, researchers will find it difficult to precisely determine a threshold.

Above the threshold, whether precisely known or not, there is a proportionate reduction in yield with increases in pest densities. Eventually, a minimum yield could be reached in which further increases in pests do not produce additional yield losses. More complicated aspects of the yield/pest relationship are discussed by Teng (16). Ideally, one attempt to maintain pest density below the threshold level, provided that the threshold exists and is known. In practice, such a feat may require more pesticides than are economically feasible. If one cannot eliminate losses in quantity or quality of yield, then an economically optimal yield should be strived for. Such an optimum is achieved by maximizing the difference between cost of control and price of the harvested crop. Those interested in this topic should read Main (18) and Shoemaker (19).

Conclusions

There are many complex interactions among pests and their biotic and abiotic environment. Knowledge of these interactions, pest population dynamics, yield losses, and pest density and aggregation should improve our ability to properly use pesticides in the ecosystem. Despite their world-wide importance, much of this information on pests in the ecosystem still needs to be determined. Agricultural researchers, mainly at land grant schools and experiment stations, are continuing a long tradition of working in these areas. At present, our knowledge of the pest as part of the ecosystem is substantial only for a relatively few species. In an era when large percentages of new research dollars are being spent on biotechnology, researchers, unfortunately, may have a difficult time in acquiring the supplies, equipment, and personnel needed to carry out large field studies. Administrators must be made aware of the importance of this research.

One benefit of additional research is the expansion of our knowledge and understanding of such factors as damage thresholds, pest population dynamics, and how pests interact with other organisms in the ecosystem and react to changes in the environment. The other benefit of this work is that agricultural researchers will have the data to better educate others. Obviously, graduate and undergraduate students, as well as co-workers, will be the first to benefit from a better understanding of pests in the ecosystem. Some of these studies eventually will have positions with pesticide producers or IPM programs where they can apply their knowledge and also educate many others. Fortunately, this education is currently being carried on and will continue and improve only if universities, the federal and state governments, and industry continue to support researchers and teachers concerned about understanding pests as part of the ecoystem.

Literature Cited

1. James, W. C.; Teng, P. S.   In "Applied Biology. Vol IV";
    Coaker, T. H., Ed.; Academic: New York, 1979; pp. 201-267.
2. Zadoks, J.; Scheen, R. D.   "Epidemiology and Plant Disease
    Management"; Oxford: New York, 1979; Chap. 8.
3. McEwen, F. L.   BioScience 1978, 18, 773-777.
4. Schaub, J. R.   In "Agricultural Chemicals of the Future";
    Hilton, J. L., Ed.; Rowman and Allanheld: Totowar, 1985.
5. Pimentel, D.; Krummel, J.; Gallahan, D.; Hough, J.; Merrill,
    A,; Schreiner, J.; Vittum, P.; Koziol, F.; Back, E.; Yen, D.;
    Fiance, S.   BioScience 1978, 28, 772-784.
6. "Integrated Pest Management"; Council of Environmental Quality,
    1972.
7. Stern, V. M.,; Smith, R. F.; van den Bosch, R.; Hagen, K. S.
    Hilgardia. 1959, 29, 81-101.
8. Carlson, G. A.   In "Tar Heel Economit"; Agriculture Extension
    Service, North Carolina State University, Raleigh.
9. Ruesink, W. G.   In "Sampling Methods in Soybean Entomology";
    Kogan, M.; Herzog, D. C., Eds.; Springer-Verlag: New York,
    1980; Chap. 3.
10. Campbell, C. L.; Noe, J. P.   Annu. Rev. Phytopathol. 1985, 23,
    129-148.
11. Tingery, W. M.   In "The Leafhoppers and Planthoppers"; Nault,
    L. R.; Rodriguez, J. G., Eds.; Wiley: New York, 1985; Chap. 9.
12. van Emden, H. F.   Proc. Symposium IX Int. Congr. Plant Prot.,
    1981, pp. 5-7.
13. Kranz, J.   "Epidemics of Plant Diseases"; Springer-Verlag:
    Berlin, 1974; Chap. 1-5.
14. Southwood, T. R. E.   "Ecological Methods"; Chapman and Hall:
    London, 1978; Chap. 9-12.
15. Boote, K. J.; Jones, J. W.; Mishoe, J. W.; Berger, R. D.
    Phytopathology 1983, 73, 1581-1587.
16. Teng, P. S.   In "Advances in Plant Pathology, Vol. 3,
    Mathematical Modelling of Crop Disease"; Gilligan, C. A., Ed.;
    Academic: London, 1985; Chap. 8.
17. Cousens, R.   Ann. Appl. Biol., 1985, 107, 239-252.
18. Main, C. E.   In "Plant Disease, An Advanced Treatise, Vol. I,
    How Disease is Managed"; Horsfall, J. G.; Cowling, E. B., Eds;
    Academic: New York, 1977; Chap. 4.
19. Shoemaker, C.   In "Modeling for Pest Management"; Tummala, R.
    L.; Hayes, D. L.; Croft, B. A., Eds.; Michigan State
    University: East Lansing, 1976; pp. 32-39.

RECEIVED October 7, 1986

CHEMICALS

Chapter 8

Principles Governing Environmental Mobility and Fate

James N. Seiber

Department of Environmental Toxicology, University of California, Davis, CA 95616

During the past several years, much attention has been
devoted to understanding the physical and chemical
properties, processes, and principles governing the
environmental behaviour and fate of chemicals. The goal
is to be able to predict how chemicals behave before
release occurs and to use this capability in the design
and regulation of chemicals proposed for use in pest
control and other environmental applications. This
effort has included improving the data base of key
physical and chemical properties, understanding the
processes which underly movement to air, biota, and
groundwater, and developing models for predicting
mobility and persistence. The modelling approach will
be illustrated with examples of pesticide volatilization
from water and the fate of pesticides in aquatic field
use situations. The role of field experiments in
validating predictive models will also be discussed.

Predicting how chemicals behave in the environment is a major task
facing science today. Society is no longer satisfied to know that
we can provide answers on where chemicals go and how long they
persist by conducting analyses of environmental samples after use
occurs. Rather, it demands premarket or preuse tests which can
lead to prediction, with a high degree of certainty, that the
chemical in question will not pose adverse environmental risks.
Such processes as food chain accumulation, contamination of surface
or groundwaters, undue persistence in soil or water, and movement
to sensitive environments through the air are of particular
concern. Fulfilling these expectations for premarket environmental
safety testing is a large order; it requires that much information
be available on physicochemical properties, the environmental
compartments available to the chemical in its zone of use,
processes which can transfer the chemical between compartments and
transform the chemical within each compartment, and those extrinsic
properties of the environment which influence both the course and

0097-6156/87/0336-0088$06.00/0
© 1987 American Chemical Society

rate of such processes. Given the complexity and heterogeneity of the environment--functions of both location from one environment to another and time within any given environment--it is presently not possible to provide quantitatively accurate prediction. Yet the demands of society, ever more frequently contained in regulations, require that science do the best job possible in this area. The subject of this chapter is the measurement of the key physicochemical properties which govern fate, and the use of these properties for predicting the environmental behaviour of pesticides.

Our ability to identify and measure the key physicochemical properties which influence behaviour and fate has improved considerably (1). There exist guidelines and, in some cases, detailed directions for determining such physical properties as water solubility (S), octanol-water partition coefficient (K_{ow}), bioconcentration factor (BCF), vapor pressure (P_{vp}), Henry's law constant (H), and soil sorption coefficient (K_d or K_{oc}), and the rates of such chemical processes as hydrolysis, photolysis, oxidation, metabolism by plants and animals, and biodegradation (2-4). As illustrated in Figure 1, this information along with several parameters which describe the "environment" into which the chemical is to be placed (a pond in the example) provide the starting point for making predictions on intercompartmental distribution and persistence -- the first step in defining the environmental fate for a given chemical or group of chemicals.

The second step often involves the use of physical or mathematical/computer models. Combining the "benchmark" properties (first step) with data from models (second step) allows one to draw a profile of expected behaviour. This information can be very useful to those developing chemicals for eventual release to the environment or proposing new uses for existing chemicals (5).

Certainly, if one is to rely on models, a third step must occur which involves validating the model predictions by comparing them with results from field studies. It is model validation that perhaps is in most need of immediate attention, particularly if one is aiming to regulate based upon model information as appears to be the case for EPA and several state regulatory agencies. Unfortunately, the ability of models to provide numbers may have engendered the notion that field tests are no longer needed, a notion that must be dispelled if we are to improve predictive capability to the point of regulatory reliability.

Physicochemical Properties

Polarity. A basic concept underlying virtually all physical properties, and the associated distributions involving them, is that of molecular polarity. Strictly speaking, polarity refers to the unevenness of charge in a molecule. Water is considered to be polar because it is relatively negative in the region of the oxygen atom and positive in the region of the two hydrogen atoms in the non-linear structure. The relatively high dipole moment (1.85 deByes), measured by observing the extent to which water molecules align themselves when placed between the plates of a charged condensor, and the high dielectric constant (80), measured as the ability of water to act as an insulator when placed in an electric

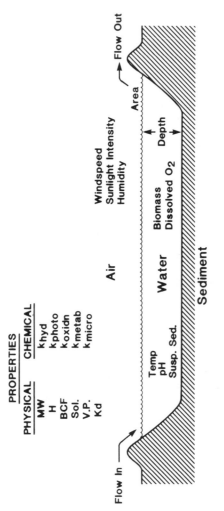

Figure 1. Intrinsic and extrinsic properties governing the distribution and fate of a chemical in a pond environment.

field, confirm this inherent polar character. Nitrobenzene is
similarly considered to be a polar aromatic compound, with a dipole
moment of 4.21 deByes and a dielectric constant of 35.7. These
values are reasonable based upon the strong polarizing effect of
the nitro substituent. We have no problem ranking nitrobenzene,
chlorobenzene (μ = 1.7 deByes, D = 5.7) and toluene (μ = 0.37
deByes, D = 2.4) in a polarity series based upon this part of the
polarity concept, and their water solubilites fall roughly in the
order expected based upon it. It is even posible to do some
ranking in simple structural series using the dipole moment contributions for various substituent groups (nitro, amino, nitrile,
etc.). Useful generalizations among otherwise similar compounds
are that symmetrical ones (eg, carbon tetrachloride) are less polar
than unsymmetrical ones (eg, chloroform), and that compounds
containing oxygen, nitrogen, and sulfur (eg organophosphate and
carbamate esters) are more polar than hydrocarbons and chlorinated
hydrocarbons. The total interaction of solute with solvent (or
with solid surface in a heterogenous environment) involves several
kinds of forces: Polar interactions (dipole-dipole, dipole-induced
dipole), hydrogen bonding, and the dispersion interactions which
exist between every pair of adjacent molecules. The latter,
referred to as London or van der Waal's forces, explain the ability
of even apolar substances to associate in condensed phases.
    A polarity ranking is not possible based only on dielectric
constant and dipole moment because they do not take into account H-
bonding; thus, polarity series are often constructed empirically,
using such factors as the solvent strength parameter obtained from
the observed ability of various solvents to elute solutes from
aluminum oxide absorbent. However, for environmental chemicals, a
numerical index for polarity does not exist; only the consequences
of polarity, as reflected in measureable properties such as water
solubility and octanol-water partition coefficient, are available
for fate predictions.

Water Solubility (S). The measurement of water solubility is
relatively straightforward for most organic compounds, involving
observation of the amount dissolved in water when an excess of the
chemical is allowed to reach equilibrium with water at constant
temperature. Centrifugation or filtration removes suspended
material from the solution prior to measurement. Experimental
variations on this basic method can produce rather large
discrepancies (6). While precision appears to be lowest with
hydrophobic compounds of very low solubility, a recent re-
measurement revealed discrepancies with literature values of up to
a factor of two for several pesticides of moderate water solubility
and a factor of 100 for two of them (ronnel and bromophos) (7). A
recently introduced column method offers the potential of
generating solubility data of good precision and accuracy much
faster than possible with the conventional method (8). This may
stimulate an effort at re-measuring all pesticides under identical
conditions.
    Water solubility is influenced by temperature (T), and the
direction generally is toward an increase in solubility with an
increase in temperature. A rule of thumb is that solubility of

solids and liquids increases by a factor of 2 for a 14° rise in temperature from 10-24°C. However, there are exceptions to this. The solubility of thiolcarbamates decreases with increasing temperature (9), an effect ascribed to an increase in resonance contribution of the uncharged -S-C(0)-N= form at higher temperatures over the -S-C(O$^-$)=N$^+$= form which predominates at lower temperatures (10). Unfortunately, solubilities in the literature are usually given at just a single temperature so that there is no basis for judging whether a regular or inverse relationship exists between S and T for a given chemical. Furthermore, the temperature may not always be specified in the literature citation, leaving a large potential for error when using such values in fate calculations.

Among organophosphates paraoxon has a water solubility of 3640 ppm compared with only 12.4 ppm for parathion (7), reflecting the much greater polarizing effect of the P=O moiety when contrasted with P=S. Similarly, phorate sulfoxide (>8000 ppm) is much more soluble than phorate (17.9 ppm) because of the presence of the polarizing S → O group in the former. It is thus not possible to estimate the water solubility of a compound based upon the value for a close analog. The effect of very small changes in structure may also help to explain some of the discrepancies in reported solubilities in the literature, where a small contamination with an analogue of much higher solubility than the compound being subjected to measurement can produce a large error in the measured value.

The water solubility of the supercooled liquid exceeds that of the solid for a given chemical above its melting temperature (t_m). An approximate formula for converting from one to the other is (11):

$$\text{Log S}_{solid} = \text{Log S}_{liquid} - 0.0095 (t_m - 25)$$

This may be an important correction in environmental fate calculations because the liquid form, rather than the solid (upon which solubility determinations are usually based), may be the state of interest in environmental processes. Obviously, the correction becomes larger for compounds of higher melting point.

Considering the above factors as well as water pH and water purity, it is clear that reported water solubilities, particularly those done at a single temperature with no indication of replication or of solute and solvent purity, must be assigned a fairly large uncertainty (at least ± 100%) when used for calculating distribution coefficients or other environmental fate parameters. When water solubility is not reported in the literature, it may be estimated from the octanol-water partition coefficient (K_{ow}) or from structural parameters (11); in either case an even larger uncertainty exists in the value.

Octanol-Water Partition Coefficient (K_{ow}). This partition coefficient is perhaps the most used distribution constant in environmental chemistry, underlying calculations of bioconcentration and bioaccumulation, several structure-activity relationships, and the choice of solvent conditions for extractions.

Partition coefficient has been studied in great detail and compilations are available in the literature (12). The laboratory measurement of K_{ow} is fairly straightforward, although again error can creep in due to such things as failure to use equilibrated solvents, non-constant temperature, and inaccuracy of the measuring technique, contributing to a fairly large uncertainty in literature values. A not atypical case is that for methyl parathion, where the literature provides at least 4 values of log K_{ow} (11):

log K_{ow}	K_{ow}
2.04	109.6
2.99	977.2
1.91	81.3
3.22	1659.6

In light of this example, literature values of K_{ow} must be given a fairly broad latitude (± 1 order of magnitude) unless they have been confirmed by more than one laboratory or by calculation from structure.

The relatively new area of property estimation has been perhaps best developed for K_{ow}. The methods of estimating log K_{ow} include:

a. Estimation from Reverse Phase - HPLC retentions
b. Estimation from water solubility
c. Estimation from structure *via* fragment constant method

Correlation with reversed phase HPLC retention data is attractive as a rapid estimation method because the sample requirements of HPLC in terms of purity and quantity are not stringent. A popular estimation method is from water solubility (S) data, given a log-log regression between S and K_{ow} for a series of compounds. An example of a regression equation applicable to mixed classes of chemicals (11) is:

$$Log\ S = 1.37\ log\ K_{ow} + 7.26$$

where S is expressed as μmol/L. Forty-one compounds, ranging from K_{ow} = 8 to 10^6, were used in the regression with a correlation coefficient (r^2) of 0.903. Other equations might be more apt for specific types of organic chemicals (11). The advantage of using the solubility correlation to obtain K_{ow} is that no chemical is required, and one only needs a literature value of S. The disadvantage is that it is just an estimation, and there is no way of assessing the accuracy of it given the uncertainty in literature values of S described above.

Related to the above is the intriguing possibility that physical properties can be calculated knowing only molecular structure, completely obviating the need for a sample of the substance or any prior laboratory work with it. For K_{ow}, the calculation from structure uses the fragment constant approach (13) or early versions of it (12). Briefly, the method employs empirically derived atomic or group fragment constants (F) and structural factors (f):

$$\log K_{ow} = \text{sum of fragments (F) and factors (f)}.$$

Values for F and f are compiled in tables (11, 13). The calculation becomes more tedious (and uncertainty in the result increases) for more complex structures. A computer program has been developed to aid in the calculation (C. Hansch and A. Leo, personal communication).

Bioconcentration Factor. The bioconcentration factor (BCF) is defined as the ratio of the concentration of a chemical in an organism to the concentration in the surrounding medium. While BCF is used most commonly as a measure of direct partitioning of chemical from water to fish, it also has some applicability to terrestrial species (plants and animals) in contact with contaminated soil or water (14).

Some confusion exists in the literature regarding the term "bioconcentration" which, as defined above, implies uptake across membranes from the medium (usually water), and "biomagnification", "bioaccumulation", and "ecological magnification". In the latter three, dietary transfer of chemical can occur along with direct partitioning. The major experimental distinction is that bioconcentration experiments are run such that no dietary intake is involved, while bioaccumulation experiments include contributions from both direct partitioning and dietary intake. In biomagnification, the use of an intact food chain involving two or more trophic levels is implied.

The measurement of bioconcentration is difficult because the water concentration must remain constant during the run and contact must be maintained until equilibrium is reached in the organism. Equilibration, signalled by a plateau in the concentration vs time plot, may take several days. This entails, particularly in the case of relatively hydrophobic compounds of low water solubility, dosing in a flow-through chamber at levels well below the toxic threshold. A complete experiment involves analysis of samples during the exposure, or "uptake phase", and also following transfer to a clean environment where release (depuration) occurs. Both the parent chemical, from which the bioconcentration factor is calculated, and known metabolites are analyzed (15).

Experimental variables include, in addition to those implied above, temperature and species of test organism. The species-to-species variation alone contributes a variability of ± 50% for the same chemical. Another variable is the type of tissue sampled. When account is taken of all error sources, values differing by much less than 1 order of magnitude may not have biological significance (15). This still leaves room for a reasonable scale as BCF for most organic chemicals fall over a wide range, from about 1 (hydrophilic compounds) to over 1,000,000 (hydrophobic chemicals).

    If BCF is not available from experimental measurements, it can
be estimated via correlation equations from water solubility (S),
octanol-water partition coefficient (K_{ow}) or soil adsorption
coefficient (K_{oc}).   Of the three, correlations from K_{ow} are
considered the most reliable because they are currently based on
the largest body of bioassay data and because K_{ow} measurements
involve a water-lipophilic phase partitioning which bears obvious
similarity to water-to-fish partitioning.   One recommended
correlation equation is (11):

$$\log \text{BCF} = 0.76 \log K_{ow} -0.23$$

It is based on data from several investigators using a variety of
fish species and 84 organic chemicals.   The log-log plot of this
correlation shows substantial scatter, underscoring the order-of-
magnitude accuracy expected in results from the use of the
correlation.

Volatility.   For vapor pressure, the fundamental property governing
condensed phase-vapor phase distributions, the experimental
measurement can be quite tedious and prone to several sources of
error particularly for compounds of low volatility.   Of the
available methods, gas saturation appears to be the most convenient
and accurate (16), while estimation based upon GC retention data
promises a more rapid (though perhaps less accurate) method worth
further development (17).
    Henry's law constant, the air-water distribution ratio, is
needed when computing either the direction of equilibrium or the
rate of volatilization from water.   It may be measured experimentally or calculated as the ratio of vapor pressure to water
solubility (18).

Rate Constants.   Distribution coefficients of the type mentioned
above tell the direction of transfer but not the rate of transfer
or overall dissipation.   As noted, the data for distribution
coefficients are imprecise and frequently difficult or impossible
to find in the literature, and their estimation techniques in need
of further improvement.   The situation is even less satisfactory
for rate constants, with the possible exception of rate of
volatilization from water (Table I).   Linear Free Energy
Relationships (LFER) offer potential as estimation techniques for
reaction rate constants, with examples being provided by estimation
of the second order rate constant for hydrolysis of organophosphate
esters from the pKa of the leaving group's conjugate acid or of
benzoic esters from Hammett sigma-rho values (11).   There are
currently just a few LFER's available, for just a few classes of
chemicals and reaction types, and the data base upon which they
have been built is fairly small particularly for pesticides.   This
is definitely an area in need of a greatly expanded effort.

Table I. Availability of Rate Constant Data from
Literature Sources and from Estimation Techniques

Rate Constant	Literature Data Base of Experimental Values	Estimation Techniques
Volatilization	Good	Good (from H)
Hydrolysis	Fair	Fair (LFER, k $\underline{vs}$ pH)
Photolysis	Fair	Fair (UV-Vis ε $\underline{vs}$ λ, solar irradiation data)
Uptake by fish	Poor	Fair (LRE from K_{ow})
Excretion by fish	Poor	Poor (LRE from K_{ow})
Uptake by soil	Fair	Not available
Desorption from soil	Poor	Not available
Biodegradation	Poor	Qualitative only

LFER = Linear Free Energy Relationship
LRE = Linear Regression Estimation

Environmental Relevance

The availability of reliable measurements or estimates of water solubility, octanol-water partition coefficient, bioconcentration factor, rate constants and the like allows one to make qualitative judgements or, through the use of mathematical simulation models such as EPA's EXAMS (19), quantitative calculations of environmental distribution and persistence. In the qualitative use, Swann and coworkers (20) classified chemical mobility in soil based upon reversed-phase HPLC retention data which in turn is related to S. The approximate water solubility equivalents in this first-estimate classification, with chemical examples, are in Table II. This classification holds for chemicals whose primary adsorption in soil is to organic matter, and excludes those chemicals (such as paraquat) which bind ionically to the soil mineral fraction. A recent tabulation of pesticides found in groundwater had 11 entries, 8 of which represented compounds with water solubilities in excess of 200 ppm with the remaining three falling in the range of 3.5 to 52 ppm (21).

Table II. Relationship Between Soil Mobility (Leaching)
and Water Solubility (20)

Mobility Class	Water Solubility	Examples
Very High	>10^6-3000 ppm	Aldicarb (5730 ppm)
High	3000-300	Bromacil (815 ppm)
Medium	300-30	Carbofuran (257 ppm)
Low	30-2	Simazine (3 ppm)
Slight	2-0.5	Ethion (1.1 ppm)
Immobile	<0.5	DDT (0.0023 ppm)

Of course, water solubility alone is not an adequate criterion for
soil movement, and must be tempered with a knowledge of soil
sorption, volatility, and chemical reactivity, competing processes
that can remove a chemical from the soil water phase, and the
method of application to soil.   An attempt to build in the
important factors which govern leaching (as well as a similar
approach to volatilization from soil) has been described (22).
Their "leaching index" can be calculated from the simple ratio:

$$\text{LEACH} = \frac{s \cdot t^{1/2}}{P_{vp} \cdot K_d}$$

Simple indices such as this can be quite useful for ranking
chemicals according to inherent leaching potential, but fall short
of direct environmental relevance because they make no account of
the soil and groundwater characteristics which exist in a given use
zone.   To factor in the latter, Aller et al. (23) developed a
weighting scheme ("DRASTIC") which combines seven hydrogeological
parameters into a score which can serve as an indicator of relative
groundwater contamination potential for a given region of the
country. The parameters are:

> Depth to groundwater
> Recharge rate
> Aquifer media
> Soil media
> Topography
> Impact of the vadose zone
> Conductivity of the aquifer

The results of both the LEACH and DRASTIC calculation can provide
important leads for selecting monitoring sites, as well as serving
to flag potentially troublesome pesticide use situation (24).   The
development of "triggers" by EPA, which take into account many of
these same groundwater contamination characteristics, indicates
that regulation based upon physicochemical properties is in the
offing.   In fact, California recently passed a groundwater
contamination prevention act (the "Connelly Bill", 25) which
mandates the setting of numerical standards for those physicochem-

ical properties involved in downward movement of chemicals through soil, with a relatively short time deadline for defining the standards. This is a relatively new development, breaking new ground in the regulatory process.

Examples of Prediction: Air-Water Distributions Involving Pesticides

The atmosphere represents an important environmental compartment for receiving and distributing residues of organic chemicals. Pesticides, for example, may enter the atmosphere during application to soil, crops, and forests by the process of drift and by volatilization of residual deposits after application (26). The concentration and form of airborne residues are of concern from the viewpoint of human exposures and for the possible damage they might cause to sensitive plants and animals downwind from the treated areas. There is also considerable evidence that the atmosphere may be an important medium for moving pesticides through the environment far from the original sites of application, and for breakdown processes.

While these points are now understood qualitatively, there is a general lack of quantitative information on specific processes for specific pesticides. The lack of such information has hampered the development of capability for predicting the relative role of atmospheric processes in overall pesticide environmental fate, and specifically of equations correlating atmospheric processes with pesticide physicochemical properties and environmental variables.

As part of a long-term study of pesticide residue dynamics in the atmosphere, we gathered and analyzed environmental samples from two situations and then compared the experimental data with results predicted by equations or models which fit the situations.

Volatilization from Flooded Fields. One instance involved the measurement of volatilization flux of pesticides from flooded rice fields in California's Central Valley. We obtained quantitative information on how much pesticide is lost to the air by post-application volatilization, at what rate the loss occurs, and what factors control it for individual chemicals.

The methodology for measuring rate of volatilization, or flux from soil, water, or crop surfaces, has been summarized by Taylor (27). In practice, one or more multiple sampling towers is placed near the center of a field or study plot, and air samples are collected at intervals for several days after application. The resulting concentrations are used to calculate flux for each sampling period, and the data from several sampling periods are then integrated to give the rate and amount volatilized. This "aerodynamic" method may be supplemented by analyzing for loss of chemical from condensed media (soil, water, foliage); the two methods should give results which are equivalent for stable chemicals or results which differ by the amount of breakdown occurring in the condensed media. While the aerodynamic method has found widespread use for measuring pesticide flux above bare soil and field crop canopies, it had not been used above water surfaces probably because of the technical difficulty in maintaining samplers at constant height and in changing sampler contents frequently without disturbing the water body. These difficulties were circumvented by making measurements in a shallow, flooded rice field with a narrow wooden pier constructed as a pathway to the air sampling and meteorological masts.

An available aquatic fate computer model, EXAMS (Exposure Assessment Modelling System, 19), provided predictions for comparison with the field-measured volatilization flux. EXAMS inputs include:

Chemical	Environment
Molecular weight	Water depth
Water solubility	Water surface area
Vapor pressure	Water temperature
	O_2 exchange constant

These parameters were known or measurable for the chemicals studied. The EXAMS model was previously used successfully for modelling volatilization processes from waste ponds (28), a somewhat similar application.

The EXAMS program is an interactive system that allows a user to specify and store the properties of chemicals and ecosystems, modify the characteristics of either using simple English-like commands, and conduct rapid evaluations and sensitivity analyses of a chemical's probable aquatic fate (19). Starting from a description of the chemistry of a toxicant, and the relevant transport, physical and chemical characteristics of the ecosystem,- EXAMS computes:

1) Exposure: the ultimate (steady-state) environmental concentrations resulting from a specified pattern of pollutant loadings,
2) Fate: the distribution of the chemical in the system and the fraction of the loadings consumed by each transport and transformation process (volatilization, in the case studied here),
3) Persistence: the time required for effective purification of the system (via export/transformation processes) once the pollutant loadings terminate.

Field-measured herbicide volatilization from flooded rice fields, results from a laboratory simulation chamber (28), and predictions from the EXAMS computer program, led to several conclusions (29). Both the laboratory chamber and EXAMS computer model showed good potential for predicting the volatilization flux of pesticides applied to flooded fields. EXAMS calculations agreed fairly well, and laboratory chamber measurements agreed very well with field results for the thiocarbamate herbicides, thiobencarb and molinate (Table III). For MCPA, neither EXAMS nor the laboratory chamber gave flux values approaching those observed in the field, but in this case the major sources of volatilized residue were deposits on dry foliage and soil surfaces rather than from solution in water. The MCPA case also showed the potential

Table III.   Summary of Normalized Flux Values For Chemicals
in Flooded Rice Fields (29)

		Flux ($ng/cm^2 \cdot hr \cdot ppm$)			
				Rice Field	
Chemical	H ($atm \cdot m^3/mole$)	EXAMS	Laboratory Chamber	Day 1	$\bar{x}$ Days 1-3
MCPA (acid)	1.0×10^{-9}	8.1×10^{-3} (pH 3.5)	4.1×10^{-3} (pH 3.5)	2.8	1.9
MCPA-DMA (salt)	$<10^{-13}$	0.0000	----	----	----
4-Chloro-o-cresol	1.1×10^{-6}	---	---	330	243
Thioben-carb	1.7×10^{-7}	4.5	23.8	23	23
Molinate	9.6×10^{-7}	51.5	62.8	66	47

importance of a relatively minor contaminant/conversion product 4-chloro-o-cresol having much higher volatility than the parent pesticide as a contributor to airborne residues.

Although the objective was to compare results from EXAMS, the laboratory chamber, and field in this evaluation, the findings do allow for assessment of the relative importance of volatilization as a fate process for the three herbicides studied, showing clearly that volatilization rate decreases in the order molinate > thiobencarb > MCPA which is in agreement with the prediction based upon Henry's law constant. The ability of the model to generate data supporting the field measurement bodes well for the further use of such models in the future. In the case of volatilization, which is so difficult to measure experimentally, the availability of a predictive model would be a welcome development.

While this study showed the potential of EXAMS for forecasting volatilization over fairly broad time intervals, a more refined study was needed to supply experimental flux data to test the capability of EXAMS to model variations in flux with time of day, windspeed, and temperature. This was conducted for molinate using the same basic field design as before but with more sampling intervals and heights, and better micrometeorological equipment. EXAMS was provided with inputs of temperature, windspeed, and with molinate water solubility and vapor pressure corresponding to the temperatures at each sampling interval. The results (30) showed that EXAMS correctly forecast the flux maxima and minima measured by the aerodynamic method. The experimental measurements from the field were somewhat lower than predicted by EXAMS and by the loss in field water concentrations of the herbicide.

Overall, EXAMS appeared to be quite promising as a predictive tool for estimating volatilization loss from flooded rice fields. It may be useful for estimating loss from other dissipation routes as well, a capability of the model not tested in these experiments, and for estimating overall dissipation in water from all routes of loss. This capability could be quite useful for calculating the effect of water holding intervals on the concentration of herbicides in rice field effluent reaching public waterways--a subject of much interest in the extensive rice growing regions of California's Sacramento Valley (31)--and, more generally, the concentration of virtually any organic pollutant in bodies of water at various times following contamination.

Pesticides in Fogwater. A second set of experiments dealt with the fate of pesticide in the atmosphere, and more specifically with the distribution between vapor and atmospheric moisture in the form of fog. The ways by which residues can be removed from the air include dry deposition, that is, by impaction of particles or direct air-surface exchange of vapors, and wet deposition following entrainment of particles or dissolution of vapors in fog, snow, and rainwater (32). Hundreds of organic chemicals have been identified in rainwater (33, 34) and a smaller number in fogwater (35), but aside from the more persistent halogenated materials and herbicides in rainwater (36, 37), practically no measurements have been made of other pesticides.

In 1983-84, a collection system designed by USDA-ARS personnel Glotfelty and Liljedahl at Beltsville, MD, was used to collect fogwater from Maryland. In this collaborative project, an analytical method (adapted from 38) was developed for ppt concentrations of representative pesticides, and positive findings were made of five organophosphates (including diazinon, malathion and methyl parathion), two triazines (atrazine and simazine), an organochlorine (DDT), and several phthalate esters and polycyclic aromatic hydrocarbons. In 1984-85, the fog sampler was brought to California for sampling fog from the Central Valley. These "tule fogs" may linger for several days in the December-March season and, in the process, entrain airborne dusts and partition chemical vapors. A total of 16 phosphorus- and nitrogen-containing compounds were measureably present in fogwater collected at the Kearney Agricultural Center near Fresno (Table IV) and 16 in another sample collected near Corcoran in the cotton-growing region of Kings County; similar findings were obtained for samples from other parts of the Central Valley (39). The concentrations of several of these chemicals--notably, p-nitrophenol, diazinon, ethylbenzimidazole, parathion, paraoxon, chlorpyrifos, and DEF-- were suprisingly high, extending to 30 ppb in the extreme case of p-nitrophenol. For diazinon, Central Valley fog had approximately 20 times the concentration measured in Maryland fog during the preceding winter. We were particularly intrigued by two findings:

  1. Breakdown products of parathion (paraoxon and p-nitrophenol), trifluralin (ethylbenzimidazole), and chlorpyrifos (chlorpyrifos oxon) were surprisingly significant residues.

TABLE IV. Distribution of Chemicals Between Fog Water and
Interstitial Air - Kearney Agricultural Center,
January 13, 1985 (39)

Chemical	Concentration, ng L^{-1} Fog Water	Air $(X10^3)$	Distribution Ratio $(X10^6)$ Air/Water Experimental	Literature
Diazinon	16,000	2.2	0.12	60
Parathion	12,400	3.2	0.25	9.5
Chlorpyri-fos	1,020	3.3	3.2	500
Methida-thion	840	<0.03	<0.04	0.07
DEF	250	<0.03	<0.12	320
Malathion	70	<0.03	<0.4	1.0
p-Nitro-phenol	32,800	<1.2	<0.04	0.084
Simazine	390	<0.1	<0.3	0.025
Atrazine	270	<0.2	<0.7	0.2
Paraoxon	9,000	0.21	0.023	0.25
Methida-thion Oxon	120	<0.03	<0.25	----
Diazoxon	190	<0.03	<0.16	----
Chlorpyri-fos Oxon	170	<0.03	<0.18	----
Triflura-lin	<350	1.3	>3.7	3.2×10^3
Ethylben-zimidazole	14,000	2.2	0.15	----
PCNB	<3,000	113	>40	1.8×10^3

2. Unlike the Maryland findings, the distribution of
chemicals between vapors and fogwater in California
samples differed considerably from that predicted by
Henry's Law constants, being much more enriched (100
to 1000 x) in the water than predicted.

The observation of breakdown products may reflect hydrolysis
or oxidation of residues on surfaces prior to volatilization or
conversion in the vapor phase (40). The unexpectedly high water
phase distribution is more difficult to explain and, in the present
context, shows that our ability to predict behaviour is far from
perfect. Apparently, partitioning in fog atmospheres is not
simple, and might involve a contribution from particle entrainment
or from surface-active solutes which enhance dissolution in the
water phase. The encouraging point is that pesticides were
correctly predicted to be measureably present in fogwater; this
finding may have pinpointed an environmental medium (fogwater) that
can be used advantageously to measure movement and dispersion of
pesticides, and also provide basic information on an environmental
fate pathway previously unrecognized. Although it is doubtful that

fogwater will concentrate pesticides to an extent that poses a
health hazard, it certainly should not be overlooked when assessing
the total exposure for humans, wildlife, and plants.

Conclusions

The study of the behavior and fate of chemicals in the
environment -- environmental chemodynamics -- has moved from
reliance on retrospective analytical data toward an ability to
predict quantitatively based upon properties which can be
determined in the laboratory or estimated from structure. The
important physical properties and the corresponding distribution
coefficients may be measured directly or estimated by linear
regression equations. The advantage of this approach is that it
allows one to screen for chemicals that are likely to show adverse
environmental behaviour characteristics (eg, leaching or
biomagnification) early in the development phase, and in fact to
build in desireable properties (eg, by addition of appropriate
functional groups) to prevent unwanted persistence or contamination. The disadvantages at the present are that (1) literature
values of the key properties, as well as those obtained by
regression correlations, are only approximate so that conclusions
drawn from calculations based upon them are usually rough
estimates, and (2) some environmental processes (such as leaching)
are not well enough understood presently to allow quantitative
predictions even if physical properties of high accuracy were
available.

Certainly the trend toward refining our ability to measure or
estimate the appropriate properties, to correlate these with
structure, and eventually use them to calculate environmental
behaviour and fate of existing chemicals or to design new
alternative chemicals will continue. For the present, field
testing is still needed both to point out adverse behavior not
fully understood or anticipated and to provide data to ensure that
our property-based estimations are headed in the right direction.
It is important that such validation studies be pursued vigorously
given the rush toward adopting models, and the results of models,
for regulatory purposes. Regulating pesticide uses based upon
physicochemical properties and models poses new challenges which,
if successfully met, will ensure a steady improvement in the
development and use of agrochemicals posing minimal risks to man's
environment.

Literature Cited

1. Seiber, J.N.   In "Agricultural Chemicals of the Future";
    Hilton, J.L., Ed; Rowan and Allanheld: Totowa, NJ., 1985;
    Chapt. 31.
2. "Test Protocols for Environmental Fate and Movement of
    Toxicants," Association of Official Analytical Chemists,
    Arlington, VA., 1981.
3. "Annual Book of ASTM Standards, Part 3, Method D," American
    Society for Testing and Materials, Philadelphia, 1970, p.
    1193.

4.   "Chemical Fate Test Guidelines," Office of Pesticides and
     Toxic Substances, USEPA Report EPA560/6-82-003, PB82-233008,
     Washington, D.C., 1982.
5.   Swann, R.L.; Eschenroeder, A., Eds. "Fate of Chemicals in the
     Environment - Compartmental and Multimedia Models for
     Predictions"; ACS SYMPOSIUM SERIES No. 225, American Chemical
     Society: Washington, D.C., 1983.
6.   Gunther, F.A.; Westlake, W.E.; Jaglan, P.S. Residue Rev.
     1968, 20, 1-148.
7.   Bowman, B.T.; Sans, W.K. J. Environ. Sci. Health 1979,
     B14(6), 625-34.
8.   Wasik, S.P.; Miller, M.M.; Teari, Y.B.; May, W.F.; Sonnefeld,
     W.J.; Devoe, H.; Zoller, W.H. Residue Rev. 1983, 85, 29-42.
9.   Freed, V.H.; Hague, R.; Vernetti, J. J. Agric. Food Chem.
     1967, 15, 1121-3.
10. Rummens, F.N.A.; Louman, F.J.A. J. Agric. Food Chem. 1970,
     18, 1161-4.
11. Lyman, W.J.; Reehl, W.F.; Rosenblatt, D.H. "Handbook of
     Chemical Property Estimation Methods"; McGraw-Hill: New York,
     1982.
12. Leo, A.; Hansch, C.; Elkins, D. Chem. Revs. 1971, 71, 525-
     621.
13. Hansch, C.; Leo, A.J. "Substituent Constants for Correlation
     Analysis in Chemistry and Biology"; John Wiley: New York,
     1979.
14. Kenaga, E.E. Environ. Sci. Technol. 1980, 14, 553-6.
15. Macek, K.J.; Petrocelli, S.R. In "Test Protocols for
     Environmental Fate and Movement of Toxicants", AOAC:
     Arlington, VA., 1981; pp 168-76.
16. Spencer, W.F.; Cliath, M.M. Residue Rev. 1983, 49, 1-47.
17. Kim, Y.-H.; Woodrow, J.E.; Seiber, J.N. J. Chromatogr. 1984,
     Vol. 314.
18. Mackay, D.; Shiu, W.Y.; Sutherland, R.P. Environ. Sci.
     Technol. 1979, 13, 333-7.
19. Burns, L.A.; Cline, S.M.; Lassiter, R.R. "Exposure Analysis
     Modeling System (EXAMS): User Manual and System Documen-
     tation"; U.S. Environmental Protection Agency; Environmental
     Research Laboratory: Athens, GA., 1981.
20. Swann, R.L.; Laskowski, D.A.; McCall, P.J.; Vander Kuy, K.;
     Dishburger, H.J. Residue Rev. 1983, 85, 17-28.
21. Cohen, S.Z.; Creeger, S.M.; Carsel, R.F.; Enfield, C.G.
     In "Treatment and Disposal of Pesticide Wastes"; Krueger,
     R.F.: Seiber, J.N., Eds.; ACS SYMPOSIUM SERIES No. 259,
     American Chemical Society: Washington, D.C., 1984: pp 297-326.
22. Laskowski, D.A.; Goring, C.A.I.; McCall, P.J.; Swann, R.L.
     In "Environmental Risk Analysis For Chemicals"; Conway, R.A.,
     Ed.; Van Nostrand Reinhold Co.: New York, 1982.
23. Aller, L.; Bennett, T.; Lehr, J; Petty, R. "DRASTIC: A
     Standardized System for Evaluating Groundwater Pollution
     Potential using Hydrogeologic Settings"; USEPA, Office of
     Research and Development: Washington, D.C., 1985; 384 pp.
24. Rao, P.S.C.; Hornsby, A.G.; Jessup, R.E. Soil Crop Sci.
     Florida Proc. 1985, 14, 1-8.

25. "AB 2021, California Legislature," State of California, Sacramento, CA., 1985.
26. Seiber, J.N.; Kim, Y.H.; Wehner, T.; Woodrow, J.E. In "Pesticide Chemistry: Human Welfare and the Environment"; Miyamoto, J.; Kearney, P.C., Eds.; Pergamon Press: Oxford, 1983; Vol. 4.
27. Taylor, A.W. J. Air Pollut. Control Assoc. 1978, 28, 922.
28. Sanders, P.E.; Seiber, J.N. In "Treatment and Disposal of Pesticide Wastes"; Krueger, R.F.; Seiber, J.N., Eds.; ACS SYMPOSIUM SERIES No. 259, American Chemical Society: Washington, D.C., 1984; pp 279-96.
29. Seiber, J.N.; McChesney, M.M.; Sanders, P.F.; Woodrow, J.E. Chemosphere 1986, 15, 127-38.
30. Seiber, J.N.; McChesney, M.M.; Woodrow, J.E. "Experimental validation of model-predicted volatilization rates of pesticides from water"; Paper presented at the 190th National Meeting of the American Chemical Society (AGRO 99), Chicago, IL, Sept. 8-13, 1985.
31. Cornacchia, J.W.; Cohen, D.B.; Bowes, G.W.; Schnagl, R.J.; Montoya, B.L. "Rice Herbicides: Molinate (Ordram) and thiobencarb (Bolero)"; California State Water Resources Control Board: Sacramento, CA, April, 1984; Special Projects Report No. 84-45P.
32. Galloway, J.N.; Eisenreich, S.J.; Scott, B.C., Eds. "Toxic Substances in Atmosphere Deposition: A Review and Assessment"; National Atmospheric Deposition Program, NC-141: July, 1980; 146 pp.
33. Gill, P.S.; Graedel, T.E.; Weschler, C.J. Rev. Geophys. Space Phys. 1983, 21, 903.
34. Duce, R.A.; Mohnen, V.A.; Zimmerman, P.R.; Grosjean, D.; Cautreels, W.; Charfield, R.; Jaenicke, R.; Ogren, J.A.; Pellizari, E.D.; Wallace, G.T. Rev. Geophys. Space Phys. 1983, 21, 921.
35. Munger, J.W.; Jacob, D.J.; Waldman, J.M.; Hoffman, M.R. J. Geophys. Res. 1983, 88, 5109.
36. Eisenreich, S.J.; Looney, B.B.; Thornton, J.D. Environ. Sci. Technol. 1981, 15, 30.
37. Pankow, J.F.; Isabelle, L.M.; Asher, W.E.; Kristensen, T.J.; Peterson, M.E. In "Precipitation Scavenging, Dry Deposition, and Resuspension"; Pruppacher et al., Eds.; Elsevier: New York, 1984; pp 403-14.
38. Wehner, T.A.; Woodrow, J.E.; Kim, Y.H.; Seiber, J.N. In "Identification and Analysis of Organic Pollutants in Air"; Keith, L.H., Eds.; Ann Arbor Science: Ann Arbor, MI., 1983.
39. Glotfelty, D.E.; Seiber, J.N.; Liljedahl, L.A. Submitted for publication, 1986.
40. Woodrow, J.E.; Crosby, D.G.; Seiber, J.N. Residue Rev. 1983, 85, 111-25.

RECEIVED August 20, 1986

Chapter 9

Mammalian Metabolism

H. Wyman Dorough

Department of Biological Sciences, Mississippi State University,
Mississippi State, MS 39762

While pesticide risk assessment should never become
totally a predictive process, it is essential that
our prior knowledge of xenobiotic metabolism be used
to help predict the hazards of chemicals before they
are introduced into the environment. It is important
to realize, however, that even a complete knowledge of
the metabolism of a pesticide in mammals does not necessarily alert one to the potential safety or risks that
might be associated with the use of that particular
compound. This can be accomplished only when the metabolism data can be related to toxicological significance.
Consequently, greater emphasis in the future must be
placed on defining the influence of given metabolic reactions on toxicity, while at the same time continuing the
awesome task of individual metabolite identification and
toxicity assessment.

The recent space shuttle disaster which claimed the lives of seven
American astronauts resulted in one of the most extensive and
expensive investigations ever conducted. Every phase of the space
program was included in the investigation and flaws causing, or at
least contributing to, the accident were eventually revealed that
seemingly would never occur in such a sophisticated undertaking.
This dreadful event is mentioned here just as one example where
proper caution, even amongst the best, sometimes is not practiced.
Like most after-the-fact investigations, the space shuttle incident
raises a point that is germane to all whose actions and reactions
may impact upon the safety of others. That is, it is not only
important that the right questions are being asked, it is essential
that the right questions are being asked at the right time.
One of the most important aspects of the symposium on pesticide
risk is the opportunity it provides for those in the pesticide
chemistry and toxicology field to think about whether the right
questions are being asked and if they are being asked at the right
time. Such is not always the case as evident by the numerous
incidences where caution and, indeed, most of our regulations

0097-6156/87/0336-0106$06.00/0
© 1987 American Chemical Society

concerning pesticide safety, have resulted largely from after-the-fact considerations. Fortunately, the frequency of scientists reacting to pesticide-related crisis, in lieu of sound planning towards their prevention, seems to be decreasing. As amply evidenced by the recent situation regarding pesticide-contaminated ground water, however, all is not well and it is clear that there are still many cases where the right questions have not been asked, or at least asked of the right people. Ideally, the right people would be those scientists who are experts in the area of pesticides and who, if specifically asked, could accurately predict many hazards before they occur. That this is not always done, even when the expertise exists, attests to avoidable weaknesses in the present hazard assessment system.

The present paper is concerned with pesticide metabolism and in keeping with the issue of question-asking as heretofore discussed, the question which must be addressed is as follows:

> Can and will a better understanding of mammalian metabolism of pesticides minimize pesticide risk?

On the surface a simple question seemingly deserving a simple answer, this question is very difficult and potentially exceedingly complex because of some apparent and some not so apparent associated ramifications. Since an answer would depend upon a precise and perhaps very different personal interpretations of the question, some analysis of the question is in order.

Metabolism and Risk

Understanding vs Knowledge. The term "understanding" takes on paramount importance in any attempt to relate pesticide metabolism to the issue of hazards and pesticide risk. A thorough knowledge of pesticide metabolism in mammals is essential to any fundamental understanding of the processes involved, but it is safe to say that we have a far greater knowledge than understanding of the subject. Although a monumental task and one that is far from complete, it is presently quite possible to identify essentially all metabolites of most pesticides in any given system. This in itself, however, does not mean that the risk associated with the use of the chemical involved, or its analogs, has been minimized.

When the products formed by metabolic processes are toxicologically insignificant and, when this is a known fact, the findings may be valuable in assessing pesticide risk. Contrarily, a number of pesticides yield metabolites known to be highly toxic and these materials may be taken into account in the risk assessment process. Too many times, however, the isolation and identification of pesticide metabolites tell us very little about risks that may be associated with the use of a particular chemical because the information can not be related to in vivo toxicological significance.

Time, economic, and technical limitations make the proper testing of all metabolites for all types of toxic action virtually impossible. Consequently, few metabolites, per se, are thoroughly tested and, even then, the type of toxicity assessed is usually limited to acute in vivo situations and to select batteries of

short-termed bioassays. Moreover, one must be especially mindful
that the possibility always exists that the tests utilized do not
allow expression of the specific toxic characteristics of the
compound being evaluated. Furthermore, there is the ever-present
possibility that the test system is not applicable to humans. The
conclusion must be drawn, therefore, that it is far easier to
correctly predict what metabolite will be formed than to predict
what affect its formation will have on the hazards of the pesticide
being evaluated.

Using Knowledge to Predict Risk. Before declaring the whole situation hopelessly disastrous, a brighter side of the metabolism and
risk should be considered. That is, when sufficient knowledge of
pesticide metabolism in mammals is accumulated to begin to understand the basic processes involved and predict how these processes
are likely to influence toxicity, then, we may begin to use that
knowledge in an intelligent manner as suggested by Williams (1). In
1968 this metabolism scientist wrote that "The intelligent use of
our knowledge of the biochemistry of foreign compounds should permit
us to predict what compounds are safe to use and to avoid those,
like thalidomide, which produce adverse effects". While the
intelligent use of our knowledge of pesticide biochemistry is not
likely to ever allow us to always predict which compounds are safe
and to avoid those which produce adverse effects as proposed by
Williams, great strides towards this ultimate objective have been
made in recent years.

For example, it is presently well recognized that many nitroso
compounds are highly carcinogenic chemicals (2) and to see today a
nitrosoamine or nitrosoamide introduced as a potentially new pesticide would be very unlikely. Additionally, while many pesticides in
use today undergo nitrosation, any new compound which might do so
would be viewed with great skepticism by most toxicologists should
such a chemical be proposed as a new pesticide. At a minimum, such
a compound would be marked by our knowledge and understanding of the
toxicological properties of nitroso compounds as a product to
receive special attention from a risk assessment viewpoint. Knowing
when to exercise particular caution is the most vital component of
predictive toxicology.

Along these same lines, we would never today consider a
compound like 2-AAF (2-acetylaminofluorene) as a commercial
pesticide. This compound is listed as a carcinogen by the
Environmental Protection Agency (3) and is used by many as a
positive control in the Ames mutagenicity assay. Nonetheless, this
compound was screened for its insecticidal action after its
synthesis in 1933, long before its carcinogenic properties were
known. Fortunately, its pesticidal activity did not merit commercial
development. The probability is extremely low that a compound as
highly carcinogenic as 2-AAF would escape detection as a carcinogen
today. Not only would the chemical structure alert the toxicologist
of such a possibility, cancer is the most dreaded of all diseases
and society has demanded that science develop tests which, with
great accuracy, identify carcinogenic substances before commercial
introduction. Unlike what often appears to be the case, however,
cancer is not the only possible adverse effect of human pesticide

exposure. In time, other aspects of xenobiotic toxicity to humans
must be given the same thorough treatment by medical science as has
cancer.

Special Problems: New Compounds. A major challenge presently is to
make certain that our current knowledge and understanding of
pesticide risks be effectively put to use in the development of new
materials. The synthetic pyrethroid insecticides which are rapidly
replacing many of the older products may be used to illustrate the
point. While these compounds are insecticidal at almost unbeliev-
ably low rates, their mode of action is the same or similar to that
of DDT (4). When chlorinated, as is the case with some of the more
effective ones, they indeed become chlorinated hydrocarbons and, as
such, they must be scrutinized just as carefully as if DDT itself
was the compound under consideration. One trusts that this is being
done, but it is important that the low-rate phenomenon not be
allowed to have an undue influence as the overall toxicology of
these new pesticides are being assessed.

Chemicals capable of killing insects at rates far below those
around which our present criteria for safety evaluation were based
do pose a special problem. If they are insecticidal at such low
doses, are other aspects of their toxicological properties perhaps
as equally spectacular? The problem is not so much related to those
toxicities detected using conventional testing protocols, but to
those which have not evolved as problems with chemicals that are far
less insecticidal. Chemicals which might adversely affect the mental
state or immune capacity of humans fall within this latter category.
As often the case, it is the unknown that is most frightening and
the knowledge accumulated from years of experience with the older
materials is of little comfort when attempting to estimate all
harmful effects of new chemicals proposed as pesticides.

Metabolites and Toxicity

As previously pointed out, the problems associated with using
metabolism data to assess risks are enormous. Knowing that one
cannot accurately predict the acute toxicity of a chemical to
laboratory mice, or to different strains even after the toxicity in
one strain is known, vividly demonstrates how vulnerable our
predictions may be relating to more elusive issues such as
reproductive disorders, growth and development, mental health, aging
processes, immune deficiencies, and genetic integrity of individuals
and populations. Because of the numerous potential risks associated
with each pesticide metabolite, it is difficult to comprehend a
system that would reasonably assure adequately testing. That
metabolites do indeed influence toxicity, however, can be illustrated using specific metabolites and a given type of toxic response. A
few common examples will serve to make the point.

Heptachlor and Heptachlor Epoxide. Heptachlor is a cyclodiene
insecticide introduced around the same time as DDT and, until
recently, was used extensively. It is important to the present
discussion because it was the first case where a metabolite of a
pesticide was proven to be involved in the toxic response assumed

initially to result solely from the applied chemical. The situation came to light when heptachlor-specific means of residue analysis revealed that the "toxicant" was no longer present, but that bioassays showed otherwise. Solvent extracts of soils and treated plants continued to kill houseflies and other test species long after the heptachlor had dissipated below levels detected by chemical and instrumental analysis. Subsequent analyses demonstrated that the culprit was heptachlor epoxide, a metabolite just as toxic, if not more so, as heptachlor and much more persistent in the environment. The result of this discovery was that all future risk assessments of heptachlor had to include the metabolite, and, ultimately, the epoxide was largely responsible for the demise of heptachlor as a major agricultural chemical.

Whether the discovery of heptachlor epoxide has ever saved a human life or contributed to the prevention of a "Silent Spring" will never be known for certain. That is not important. The important thing is that the discovery, and hundreds of a similar nature which followed, provided new information applicable to most chemicals that improved the validity of the risk assessment process.

Parathion and Paraoxon. Again, this represents a reaction (the sulfur oxidation of a thiophosphate pesticide) that is familiar to most in the pesticide area. Unlike heptachlor epoxide, paraoxon is not a stable compound and its actual presence in a poisoned animal was very difficult to demonstrate. The oxons of other organophosphorothioates are not so elusive. In any event, the paraoxon metabolite is an excellent example of where an understanding of metabolic processes and their potential toxicological significance alerted scientists to the likelihood that such a metabolite existed. Many years of work with similar compounds had established that the insecticidal thiophosphates required oxidation to the P=O form in order to inhibit the neurotrasmitter acetylcholinesterase, the biochemical basis of their toxic action. Paraoxon was eventually isolated in vivo and now consideration of the oxon is a vital part of the overall risk assessment of this group of pesticides.

Carbaryl and 1-Naphthol. The mode of action of the carbamate insecticides is, like the organophosphorus compounds, inhibition of acetylcholinesterase. However, no metabolic activation is required as with the latter insecticides. Carbaryl is the most widely used carbamate and, in fact, is one of the most widely used pesticides in the world. Its acute oral LD50 to rats is usually reported to be in the 400 to 600 mg/kg range. Because chemical hydrolysis of the ester linkage yields 1-naphthol, carbon dioxide, and water, metabolism via this route would be expected to yield products of little toxicological significance to mammals. Both carbon dioxide and water are obviously of no hazard and 1-naphthol has an acute oral LD50 to rats of over 2500 mg/kg. While the metabolism of carbaryl is now known to be extremely complex and to involve formation of toxic oxidative metabolites as well as the nontoxic hydrolytic products, the latter were once thought to predominate and early risk assessments were based on this assumption (5).

Metabolic Processes and Toxicity

Isolation, identification, and determination of the toxicity of all
metabolites of a pesticide may appear to be necessary to properly
assess risk, but this is not possible and perhaps not always true.
The time and resources that would be required to do so would mean
almost certain death to those chemicals presently under development,
and no such consideration could be given to the hundreds of
pesticides already on the market. This means that other methods for
estimating the toxicity of individual metabolites must be employed.
     The process would be made rather simple if given types of in
vivo chemical and biochemical reactions were known to always have
the same effect on toxicity. That is, that all ester hydrolysis
resulted in detoxification, that all sulfur oxidations were
activations, and that all conjugations rendered a compound nontoxic
and readily excretable from the body. Unfortunately, this is not the
case. However, when applied to specific types of chemicals, our
knowledge and understanding of how certain reactions likely affect
toxicity are sufficient to be extremely valuable in estimating
pesticide risk. The more that is known about a particular chemical
group in this regard, the more likely the prediction will be an
accurate one.

Overall Metabolic Processes. In general, metabolic processes which
facilitate elimination of a pesticide from the body are considered
desirable. This is based a great deal on our long history of
associating toxicity with chemicals that accumulate in the body.
Arsenic, lead, mercury and other metals substantiate these concerns
as do more modern synthetic organic chemicals such as DDT and mirex.
Because so many chemicals rapidly voided from the body are now known
to be extremely hazardous, risks and excretion rates are evaluated
very carefully. Still, storage of metabolites is not a positive
characteristic even for those compounds like DDE whose danger, if
any, as a body burden has not been established.
     Nonetheless, so long as pesticides and their metabolites remain
in the body, the potential for damage remains. Once removed, the
potential ceases and emphasis may be placed on determining the
toxicological consequences of the chemical having passed through the
body. It should be clear, therefore, that the actual metabolic
processes which are responsible for clearance of the pesticide need
not always be known in order to be useful in risk assessment. Such
information, routes and rates of excretion, is usually available
from radiotracer studies long before the number and chemical nature
of the metabolites have been defined. A thorough evaluation of the
data at this stage is critical to the proper planning of further
metabolism studies and, combined with other data, might very well
provide a basis for conducting only a fraction of the studies which
are "recommended" in registration guidelines.

Processes vs Products. While it is often stated that certain
metabolic reactions generally represent either activation or
detoxification, care must be taken to differentiate between that
which is assumed and that which is known for certain. For example,
the metabolic conversion of thiophosphates (P=S) to phosphates (P=O)

is recognized as an activation step and, as mentioned earlier for parathion, this is certainly the case for some thiophosphates. Since the thiophosphates are weak inhibitors of acetylcholinesterase, it is probably safe to say that all compounds of this type which induce death by severe inhibition of acetylcholinesterase undergo activation via sulfur oxidation. Death without this enzyme inhibition would suggest an alternate mode of action. Does this mean, then, that metabolism of all thiophosphates to phosphates effects greater risk to the organism involved? Certainly not. Many oxons produced in the body are so unstable that the reactant is destroyed before reaching the site of detoxification. The point here is that a product may be formed which, based on similar compounds and even in vitro testing, should be highly toxic but, in fact, is not.

A situation opposite to that is seen with the carbamate insecticides. As shown in Table I, data from studies in our laboratory demonstrate that carbamate insecticides most toxic to rats are also hydrolyzed at a faster rate. Ester hydrolysis in these studies was measured by quantitation of radioactive carbon dioxide in the respiratory gases of animals treated orally with the carbamate radiolabeled on the carbonyl carbon. Yet, hydrolysis of

Table I. Carbamate Ester Hydrolysis (0-24 Hour) Following Oral
Treatment (0.2 Mg/Kg) of Rats Compared with Toxicity

Insecticide	Percent of Dose	Acute Oral LD50, Mg/Kg
Aldicarb	71.5	1
Carbofuran	54.3	5
Croneton	39.9	400
Carbaryl	27.8	600

carbamate insecticides is always considered as a detoxification reaction. The reason for this obviously is the fact that the products of hydrolysis, 1-naphthol from carbaryl, for example, are always much less toxic to mammals than the parent compound. One possibility that naturally comes to mind when the in vivo data are examined is that either hydrolysis yields a more active product or that oxidation does so and it is the latter which is hydrolyzed to carbon dioxide. This is not supported by our data to date. None of the hydrolytic products tested has demonstrated any anticholinesterase activity. Moreover, hepatic enzyme induction using mirex and other inducers, and confirmed by a number of in vitro reaction, did not alter radioactive carbon dioxide production as would have been expected if hydrolysis resulted from the formation of an oxidative metabolite.

In similar studies, we have compared the toxicity of the same carbamate insecticide to several laboratory animal species that varied considerably insofar as rates of in vitro ester hydrolysis was concerned. Again, it was the more susceptible species which demonstrated the faster rates of hydrolysis (6). Neither absorption from the gut, or total metabolism as evidenced by excretion rates and metabolites excreted, was significantly different when the same carbamate was administered to either rats, mice, guinea pigs, or gerbils. At this point, it can only be surmised that the same properties that make the carbamates excellent in vivo inhibitors of acetylcholinesterase also make them excellent substrates for other esterases.

Minimizing Risk

There is little question at this point that a better understanding of mammalian metabolism of pesticides can minimize pesticide risk. There remains, however, considerable concern about the effectiveness of the use of metabolism data in this regard and of the lack of coordinated efforts to improve the design of metabolism studies to more appropriately address the issue of risk assessment.

One of the biggest obstacle to assuring that a better understanding of pesticide metabolism in mammals will be useful in minimizing pesticide risk is the absence of any designated group of experts whose mission it is to do just that. Compound-by-compound review involving, by necessity, a few regulatory scientists and some number of industry representatives is a very inefficient and potentially dangerous means of determining pesticide risk. Different chemical groups pose different problems and it is unreasonable to expect a reviewer, or a series of reviewers, to keep abreast of the latest developments within a particular field, much less to render a decision as to whether its proposed use is allowed or disallowed.

The tendency may be to play it safe and to delay a decision as long as apparent rational reasons to do so could be devised. Groups of experts would have the same tendency, but such a group would benefit from the knowledge that is available only when those who best know a subject are brought together. While imperfect, the peer review system of NSF and NIH, and the EPA,s procedure for developing Water Criteria Documents, etc., are the types of decision-making processes which instill confidence of the nature required for making pesticide risk assessments.

With pesticide registration, it would not be feasible for a scientific review panel to rule on day-to-day matters. The role of such a panel on metabolism would be to periodically examine the requirements, or guidelines, for registration and, using select chemicals then being reviewed, determine if the process is working in the most effective and efficient manner. Ideally, there would be a general metabolism panel to consider issues pertinent to the overall topic of metabolism and pesticide risk. Similar panels of experts would then be formed to deal with each major chemical group of pesticides, and an ad hoc panel to address metabolism-risk issues of miscellaneous pesticides.

The ad hoc panel might consist of, among others, one member from each chemical-group panel, and the general panel could consist of, among others, one member from each chemical-group panel. In addition to risk matters pertaining to pesticide registration, the chemical-group panels would hopefully become the national advisory body for all major problems that arise with chemicals falling within their area of expertise. The problems are just too diverse for one panel of advisors to adequately address all pesticide issues.

For some, a proposal to "panelize" the risk assessment process may be taken as a criticism of individuals presently responsible for such matters, and/or a mechanism to share the blame. The former is totally untrue and the latter only partially true. A regulatory reviewer is somewhat like a judge; that is, one who renders a decision based on that which has been deemed proper by appropriate authorities. Their role is not to make the rules. Panels of experts would serve to assist in making the rules and in lending advice to the reviewers. Thus, the views of the reviewer, backed by the panel, would become stronger. If not backed by the panel, then, the reviewer should take comfort in the fact that she/he, or they, do not stand alone but are making a decision representative of those who are most qualified to render a judgement.

Whether a reviewer, a panel member, an industry representative, or just a concerned citizen, it is essential that each recognize that there are severe risks involved in decision making. Risk assessment of pesticides is difficult and not all decisions related thereto will be flawless. Only those who are extremely secure in their roles as scientists and as caring human beings should dare take on such awesome responsibilities.

Literature Cited

1. Williams, R. T. In "The Biochemistry of Foreign Compounds";
    Parke, D. V., Ed.; Pergamon Press: London, 1968; p. ix.
2. Mirvish, S. S. J. Toxicol. Environ. Health 1977, Vol.2, 1267-77.
3. "The Merck Index", Merck: Rahway, NJ, 1983; 10th ed., p.4058.
4. Casida, J. E.; Gammon, D. W.; Glickman, A. J.; Lawrence, L. J.
    Ann. Rev. Pharmaco. Toxicol. 1983, Vol. 23, 413-38.
5. Kuhr, R. J.; Dorough, H. W. "Carbamate Insecticides: Chemistry,
    Biochemistry and Toxicology"; CRC Press: Cleveland, 1976;
    p. 146.
6. Benson, W. H.; Dorough, H. W. Pestic. Biochem. Physiol.
    1984, Vol. 21, 199-206.

RECEIVED December 17, 1986

Chapter 10

Molecular Modeling: A Tool for Designing Crop Protection Chemicals

Erich R. Vorpagel

E. I. du Pont de Nemours & Co., Agricultural Products Department,
Experimental Station, Wilmington, DE 19898

Application of molecular modeling techniques to the biorational design of selective and environmentally safe crop protection chemicals is addressed. Sulfonylurea herbicides are used as an example to illustrate the kinds of biological information that can be known with modern technologies. An example of selective inhibitor design using computer graphics is presented.

Crop protection chemicals (CPC) are an important component in the high yield production of crops. Future trends in agriculture and CPC technology will necessitate the discovery of agrichemicals with high selectivity, high mammalian and environmental safety, low use rates, and low costs.

Traditionally, agrichemicals are discovered by empirical synthesis and evaluation. Although this approach has been (and is currently) very successful, its efficiency continues to decline. For example, in 1950 about 2,000 compounds were screened to produce one product. In 1970, the ratio was 7,500 compounds screened per product and today an estimated 20,000 compounds must be screened for every new type of product introduced. The discovery and development process can take five to eight years and cost tens of millions of dollars.

These efficiency, economic, and time realities suggest that the empirical discovery process be complemented by a chemical design program based on a molecular understanding of key biological processes related to weed, disease, and insect control (i.e., biorational design). Computer-assisted molecular modeling techniques can play an important role in both understanding these biological processes and aiding in chemical design. These techniques include a wide variety of operations associated with molecules and model building. It encompasses the generation, manipulation, and representation of three-dimensional structures of molecules and associated physiochemical properties. Because of the complexity of the systems, computers are mandatory. Since all

0097-6156/87/0336-0115$06.00/0

molecular interactions are essentially electronic, any biological
activity expressed by a molecule comes from its electron density
distribution and polarizability.   Computer-assisted molecular
modeling techniques allow the researcher to model chemical
interactions at this level (for instance, the interactions between
enzyme and substrate or enzyme and inhibitor).

In this chapter, I will address how molecular modeling tools
can be applied to the biorational design of selective and
environmentally safe crop protection chemicals.   To do this
effectively, the fundamental biological processes must be
understood.   First I will discuss the kinds of biological
information that can be known with modern technologies.   The
sulfonylurea class of herbicides that we at Du Pont are
commercializing will be used as an example.   These "sulfonylurea
herbicides" are ushering in a new era of herbicide technology.   They
will be used to illustrate how biological understanding at the
molecular level can provide valuable insight into herbicide design
strategies.   Next, I will reverse the sequence and discuss how an
understanding of biochemical processes can be combined with
molecular modeling techniques to provide general principles that can
be used to design crop protection chemicals.

Sulfonylurea Herbicides

Sulfonylureas are a new class of high potency herbicides which show
excellent weed control activity at extremely low application rates
(4 - 35 grams per hectare).   Their high potency and low use rates
combined with their high mammalian safety and crop selectivity make
them extremely attractive in terms of efficacy and the environment.
We have carried out extensive studies aimed at understanding the
factors that govern the intrinsic biological activity, crop
selectivity, and soil degradation properties of sulfonylureas.
These three properties are essential for effective and
environmentally safe herbicides.

Biological Activity.   We have shown that the site of biochemical
action for sulfonylureas is the enzyme acetolactate synthase (1,2).
This enzyme catalyzes the first common step in the biosynthesis of
the essential branched chain amino acids valine and isoleucine.
Plants must synthesize these amino acids for protein synthesis and
subsequent growth.   Therefore, this is a vulnerable or critical
enzymatic pathway.   It is important to note that plants contain
these essential amino acid biosynthetic pathways (and the associated
enzymes) while mammals do not.   Mammals obtain these amino acids
from their diet.   This largely explains why sulfonylureas are so
non-toxic to mammals [LD_{50} of >5000 mg/kg in male rats (3)].
Acetolactate synthase (ALS) catalyses the reaction of two
pyruvate molecules to give acetolactate and carbon dioxide.   It also
catalyses the reaction of pyruvate and α-ketobutyrate to give
α-aceto-α-hydroxybutyrate and carbon dioxide.   The enzyme requires
three coenzymes for activity; flavin adenine dinucleotide, thiamin
pyrophosphate, and magnesium ion.   The reaction takes place in
several steps.

First a pyruvate molecule condenses with thiamin pyrophosphate at the thiazolium ring carbon with subsequent loss of carbon dioxide. Then a second pyruvate (or α-ketobutyrate) condenses followed by loss of acetolactate and regeneration of the thiazolium ring.

Studies on enzyme kinetics show that sulfonylureas act as slow tight-binding inhibitors. They appear to bind most tightly to the enzyme after binding of the first pyruvate molecule (4). Since detailed enzyme mechanistic studies are facilitated by having large quantities of pure enzyme, bacteria have been genetically engineered to over produce the ALS enzyme. Following large scale fermentation and purification procedures, large amounts of pure ALS enzyme have been obtained. Pure enzyme is also required for growing crystals which can be used to obtain the enzyme's three-dimensional structure by X-ray diffraction. In addition we have cloned the genes coding for bacterial, yeast, and plant ALS enzymes, determined their DNA base sequences and deduced the amino acid sequences of the enzymes. These are important steps in understanding the molecular architecture of enzymes and the design of new inhibitors.

An in vitro assay for intrinsic sulfonylurea activity has been developed using isolated plant enzyme (5). The I_{50} for ALS inhibition is defined as the concentration of sulfonylurea that

inhibits ALS activity by 50%. A good correlation exists between the
herbicidal activity of sulfonylureas and their ability to inhibit
acetolactate synthase (2). This in vitro assay using the target
enzyme along with the three-dimensional structure of the enzyme
should aid in the generation of a substantial data base that can be
used to design potent inhibitors.

Crop Selectivity. I_{50} values for inhibition of the ALS enzyme from
a variety of crop and weed species have been determined (2). In all
cases the highly active herbicides proved to be potent inhibitors of
plant ALS enzyme. Crop tolerance results from rapid metabolism of

Chlorsulfuron

Inactive

Glucose

the sulfonylureas by the crop but not the weeds. Our studies have
shown (6) that the biological mechanisms for metabolic inactivation
of various sulfonylureas differ from crop to crop. For example, a
major factor responsible for the selectivity of chlorsulfuron as a
post-emergence herbicide for small grains is the ability of crop
plants, such as wheat, to metabolize the herbicide to polar,
inactive products. Sensitive broadleaf plants show little or no
metabolism of chlorsulfuron. Tolerant plants such as wheat, oats,
and barley rapidly metabolize chlorsulfuron via hydroxylation at the
5-position of the phenyl ring. This intermediate is converted to
the inactive O-glycoside by a glycosyl transferase enzyme.

Other detoxification mechanisms are also known. This diversity of metabolism by different plants and for different sulfonylurea molecules is responsible for the high selectivity found in different crops. Understanding this enzymatic and chemical diversity can facilitate the design of sulfonylureas with optimal selectivity for a particular crop.

Soil Degradation. Soil residual properties are (and will continue to be) an important parameter for herbicides and other agrichemicals. Soil degradation of sulfonylureas under field conditions occurs at rates which are similar to conventional soil active herbicides (7). Structural modification of the molecules can be used to modify the rate of degradation and thereby adjust the residual properties of the product as illustrated in Figure 1. Thus, changing from the ortho-chlorophenyl moiety in chlorsulfuron to the ortho-carboxymethylphenyl moiety in Ally causes some increase in degradation rate. However, substituting a thiophene ring for the phenyl ring gives Harmony a very rapid degradation rate. Breakdown of sulfonylureas in soil is a combination of chemical hydrolysis and microbial action (8). Knowledge of these mechanisms at the molecular level is very useful in the design of crop protection chemicals with the desired soil residual properties.

Biorational Design

Sulfonylureas, as a class of herbicides were discovered as part of an empirical synthesis and screening program (9). After much structural modification and further biological evaluation, several herbicides have been commercialized by Du Pont. The site of action, mechanisms for selectivity, and soil degradation were determined subsequent to the initial leads generated from the traditional approach. As mentioned earlier, the chances of finding another crop protection compound with the same potency, selectivity, and environmental safety are slim. The remainder of this chapter will address how an understanding of biological processes at the molecular level can be combined with computer-assisted molecular modeling techniques to design new crop protection chemicals.

Identify Target. A chemical design program usually begins by identifying a critical biological pathway in the target species. The pathway must be essential for the survival of the organism so that inhibition will cause death. The pathway should not be easily circumvented or replaced by some other process. For enzymes this means that the reaction products can not be obtained from other metabolites. Ideally, there should be differences in the intrinsic bioactivity between pest and non-pest species. For example, mammals and non-pest species should not depend on the same pathway. Some known critical biological pathways for plant species are essential amino acid biosynthesis, photosynthesis, and either the synthesis or site of action of plant hormones. For example, sulfonylureas inhibit the biosynthesis of the essential amino acids valine and isoleucine. This pathway is not present in mammals, but it is present in both crop and weed species.

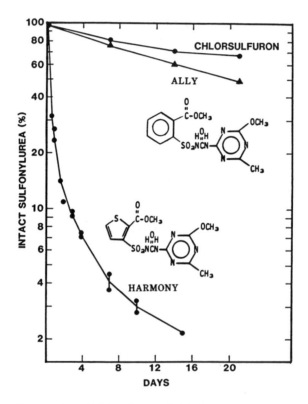

Figure 1. Sulfonylurea Soil Degradation.

Second, a key enzyme or receptor in the pathway should be identified as the target. It is best to select enzymes whose products are important for several functions in the species. Cellular response to such a metabolic blockade should also be considered (e.g., cascading effects). Often end-product limitation results in more metabolites entering the pathway. After sufficient substrate accumulation, catalysis may occur even in the presence of an inhibitor (10). However, accumulation of toxic intermediates would prevent this cellular response and lead to death. Again using sulfonylureas as an example, acetolactate synthase is a common enzyme in the pathway for two essential amino acids rather than just one. Also, inhibition of acetolactate synthase leads to high levels of α-ketobutyrate which is thought to have deleterious effects (11).

Once an enzyme or receptor has been identified, ideally it should be isolated and characterized. A three-dimensional structure is very useful and can be determined from X-ray crystallography, NMR or by some other means which may include computer-assisted molecular modeling. The mechanism for the normal catalytic activity of the enzyme should be understood. This includes knowing the natural substrates, any natural inhibitors, and any coenzyme requirements. An in vitro assay method must be developed to quantitatively test the effectiveness of potential inhibitors. At this point molecular modeling techniques can be used in the design process.

Inhibitor Design. Several approaches have been used to design enzyme inhibitors. Structural modifications of known substrates can be constructed which compete with the natural substrate. An example for the ALS enzyme would be analogues of pyruvate. Enzyme-activated irreversible inhibitors, often referred to as suicide substrates, can be designed based on some intermediate in the normal catalytic reaction. An example for the ALS enzyme would be a molecule that would react with the active site carbon atom in the thiazolium ring but could not undergo any further reaction. Transition-state analogues can be designed based on some high-energy, metastable intermediate which exists during the normal catalytic reaction. An example for the ALS enzyme would be thiamin thiazolone pyrophosphate, a known thiamin pyrophosphate analogue which has been reported in the literature (12).

Depending on the enzyme, there may be regulatory sites which can be exploited. Structural modification of known inhibitors which bind tightly can be used to permanently shut off the enzyme. For example, the amino acid valine is known to inhibit the ALS enzyme in bacteria and plants by a feedback mechanism (13). Three-dimensional molecular modeling could be used to identify a suitable cleft or pocket in the enzyme into which an inhibitor would fit and stop normal catalyic activity. Such a method has been used in drug design (14).

Because of certain proprietary considerations, a protein other than the ALS enzyme will be used to demonstrate how molecular modeling techniques can be applied to the design of crop protection chemicals. Human serum prealbumin has been used by Blaney, et al (15) to model drug-receptor interactions. Its function is to transport the hormone thyroxin. The binding site for thyroxin will be used in the following examples. Figure 2 is a computer generated representation of the three-dimensional structure of the binding site. Thyroxin is shown as a stick structure in red. A series of dots has been added to the surface of the binding site. This solvent-accessible surface is generated (16) by mathematically rolling a probe sphere over the surface of the molecule. It is displayed as a continuous envelope of dots which can be color coded to show various physiochemical properties at the surface.

Molecular modeling techniques can be used to fit novel compounds into the binding site. They need not be structurally similar to the natural substrate but the dominant physiochemical properties should be similar. Modifications can be made to the molecule to improve the fit. This will increase specificity for the target enzyme. Substituents can be added or modified so that regions in the enzyme interact favorably with parts of the inhibitor. Electrostatic interactions, hydrophobicity, and hydrogen-bonding can be included in the fitting process. By judicious choice of substituents, both in vitro and in vivo activity can be optimized. Substituents could be added or modified to improve uptake, translocation, and accumulation in the appropriate parts of the pest species provided the molecule still fits in the binding site.

Selectivity of the compound for pest and non-pest species can also be designed into the molecule with the aid of computer-assisted molecular modeling techniques. There are several ways to affect this selectivity; differential inhibition of the enzyme, differential uptake, or rapid metabolism of the inhibitor by the tolerant species. Differential inhibition of the enzyme is preferred because it is less susceptible to the metabolic states of the pest and non-pest which can be influenced by environment, stage of development, and behavior.

If the target enzymes for both pest and non-pest species have been isolated and characterized, they can be used to design compounds which are specific for the pest species. For example, Figure 3 shows the binding sites for both the pest (left) and non-pest species (right). The shape of the binding site (blue dot surface) is slightly different in the lower left part of the binding site (see arrows). This is the result of a single amino acid residue change near the binding site. This modification was made to prealbumin using computer-assisted molecular modeling techniques. There is precedent for such an amino acid change affecting enzyme inhibition (11). Yeast mutants have been isolated which are highly resistant to sulfonylureas. There is a single amino acid change in their ALS enzyme.

An inhibitor has been included in the two binding sites (red dot surface). Notice the extra space present in the pest species enzyme at the ortho-position (see arrows). This space is not present in the non-pest species. This suggests that an analogue with an appropriately-sized substituent in this position should inhibit the pest species enzyme but not the non-pest enzyme.

Figure 4 shows the two binding sites with an ortho-substituted analogue. Based on shape, this analogue does not fit in the non-pest enzyme (see arrow). The surface of the inhibitor goes

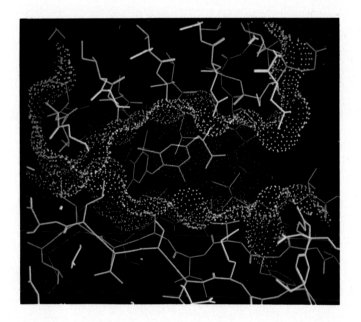

Figure 2.   Cross Section of the Thyroxin-Prealbumin Binding Surface.

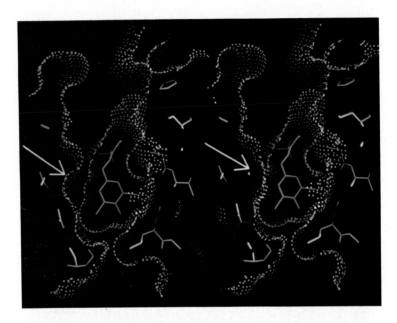

Figure 3. Cross Section of the Pest and Non-Pest Binding Sites with a Non-Selective Inhibitor.

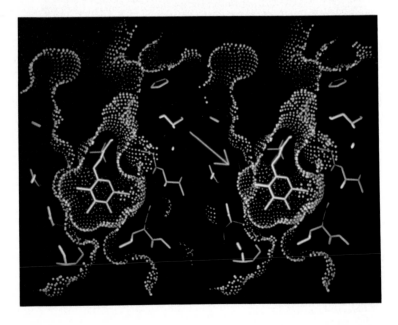

Figure 4. Cross Section of the Pest and Non-Pest Binding Sites with a Selective Inhibitor.

beyond the surface of the enzyme. This compound should show selectivity in the in vitro assay. Also, compounds with larger substituents in the ortho-position probably would not be inhibitors of either enzyme.

Ultimately, the final crop protection compound will require the best combination of potency, uptake, translocation, selectivity, metabolism, degradation, and minimal toxicity to non-pest organisms. This is the same set of criteria that all crop protection chemicals have to meet. The only real difference with the methods detailed here is that the lead compound and the decisions of how to modify its chemical structure came from a knowledge of the biochemistry. This approach may significantly decrease the number of analogues that must be made and result in a more efficient discovery of crop protection chemicals which are potent, selective and of minimal risk to the environment.

Is this approach practical? We believe so, even though this field is in its infancy. Similar approaches have been used in medicinal chemistry where X-ray structures of key enzymes are known (14,17). Their success in the design of potent inhibitors using molecular modeling techniques is encouraging. The biggest difficulty in applying these techniques to design crop protection chemicals is the limited amount of basic scientific knowledge. We need to know more about the physiology, biochemistry, and molecular biology of insects, fungi, and plants. As more is learned about these systems, computer-assisted molecular modeling will become a more useful and effective tool.

Literature Cited

1. La Rossa, R. A.; Schloss, J. V. J. Biol. Chem. 1984, **259**, 8753-8757.

2. Ray, T. B. Proc. British Crop Protection Conf.-Weeds, 1985, **3A-1**, 131-138.

3. Levitt, G.; Ploeg, H. L.; Weigel, R. C., Jr.; and Fitzgerald, D. R. J. Agric. and Food Chem. 1981, **29**, 416.

4. Schloss, J. V. In "Flavins and Flavoproteins"; Bray, R. C.; Engel, P. C.; Mayhew, S. G., Eds.; Walter de Gruyter & Co.: Berlin, 1984; pp. 737-740.

5. Ray, T. B. Plant Physiol. 1984, **75**, 827-831.

6. Sweetser, P. B.; Schow, G. S.; Hutchison, J. M. Pest. Biochem. and Physiol. 1982, **17**, 18-23.

7. Palm, H. L.; Riggleman, J. D.; Allison, D. A. Proc. British Crop Protection Conf.-Weeds, 1980, **1**, 1.

8. Joshi, M. M.; Brown, H. M.; Romesser, J. A. Weed Science 1985, **33**, 888-893.

9. Saures R. F.; Levitt, G. In "Pesticide Synthesis Through
    Rational Approaches"; Magee, P. S.; Kohn, G. K.; Menn, J. J.,
    Eds.; ACS SYMPOSIUM SERIES No. 255, American Chemical Society:
    Washington, D.C., 1984; pp. 21-28.

10. Christopherson R. I.; Duggleby, R. G. Eur. J. Biochem.
    1983, **134**, 331-335.

11. La Rossa, R. A.; Falco, S. C. Trends Biotechnol. 1984, **2**,
    158-161.

12. Gutowski J. A.; Lienhard, G. E. J. Biol. Chem. 1976, **251**,
    2863.

13. De Feliece, M.; Lago, C. T.; Squires, C. H.; Calvo, J. M. Ann.
    Microbiol. (Paris) 1982, **133A**, 251-256.

14. Goodford, P. J. J. Med. Chem. 1984, **27**, 557-564.

15. Blaney, J. M.; Jorgensen, E. C.; Connolly, M. L.; Ferrin, T.
    E.; Langridge, R.; Oatley, S. J.; Burridge, J. B.; Blake, C. C.
    F. J. Med. Chem. 1982, **25**, 785-790.

16. Connolly, M. L. Science 1983, **221**, 709-713.

17. Hopfinger, A. J. J. Med. Chem. 1985, **28**, 1133-1139.

RECEIVED August 4, 1986

POTENTIAL HAZARD

Chapter 11

Pesticide Use: The Need for Proper Protection, Application, and Disposal

W. K. Hock

The Pennsylvania State University, University Park, PA 16802

Current pesticide management practices can result in
three categories of human exposure situations--acute,
chronic high or occupational, and chronic low or
incidental. Pesticide exposure, whether direct or
via chemical trespass from treated areas, can be
reduced, if not eliminated entirely, by utilizing:
(a) adequate personal protective equipment, (b)
technologically superior application equipment and
techniques, and (c) improved and economically affordable disposal processes. New lightweight and
inexpensive protective clothing and equipment need
to be developed if applicators are expected to comply
with personal protection requirements on pesticide
labels. Research in application technology needs to
address the issue of applicator exposure as well as
that of efficacy and economics, and reliable cost
efficient disposal techniques need to be developed
for small volume pesticide users.

By design, pesticides are biologically active and, in most cases,
toxic. Thus, they pose potential risks to human beings and other
living organisms (1). As is the case with toxic substances in
general, pesticides pose several different kinds of threats to
health. These adverse effects are commonly considered as either
"acute" effects, developing quickly after exposure but of usually
short duration, or "chronic" effects, which may appear after a delay,
often years, but then persist for extended periods (2). Chronic
adverse effects occur as a result of sustained exposures but are
much more difficult to evaluate than are acute effects (Table I).
Such exposures are often classified as "chronic high" resulting from
occupational exposures, or "chronic low" occuring from low level,
incidental exposure (3). We know that there are some chronic effects
from particular chemicals, but securing the documentation for
possible long-term effects such as increased prevalence rates of
cancer, vascular disease, and organ injury is extremely difficult,
and may well be impossible to obtain (2).

0097-6156/87/0336-0128$06.00/0
© 1987 American Chemical Society

Table I. Chronic Adverse Effects of Pesticides

Chronic effects (delayed onset, or protracted, recurrent, or
  irreversible course)
A. Peripheral neuropathy
B. Effects on reproduction
C. Sensitization
D. Suspected, but generally unconfirmed effects:
    1. Effects on brain, heart, liver, kidney, lung, blood,
          reproductive organs
    2. Accelerated atherosclerosis, hypertension
    3. Carcinogenesis
    4. Teratogenesis
    5. Impaired immunity and immunopathies

Source: Ref. 2.

With an excessive, single exposure, the result will be either a
systemic pesticide poisoning or a topical lesion frequently observed
on the skin or in the eyes. Since most acute intoxications are from
the carbamate and organoposphate insecticides, the systemic manifestations are cholinergic and are due to the inhibition of acetyl
cholinesterase and the resultant accumulation of the neuro-
transmitter acetylcholine, at the synapse. Topical effects, in
contrast, either are the result of the irritant properties of the
chemicals in the formulation or have an allergenic basis for their
occurrence (3). However, topical effects are not necessarily
exclusively the result of exposure to the active ingredient in the
formulation but may result from a reaction to one or more inerts as
well.
     With chronic or sustained exposure to pesticides, the popula-
tions at risk are those who receive repetitive exposures during the
manufacture, formulation, mixing, application, or disposal of pesticides. Another type of chronic exposure is persistent residue
contact by workers in the field during the harvesting and thinning
of fruits, vegetables and other agricultural commodities. The
outcome of these repetitive exposures can result in a number of
different diseases.
     Although of considerable public concern, the chronic incidental
exposure that the general public receives from trace amounts of
pesticide residues in air, food, and water, does not usually result
in a public health crisis to the population at large. In such
instances a pesticide exposure profile is required to determine the
extent and/or severity of this incidental, involuntary pesticide
exposure (3).
     With pesticides being used currently at a rate approaching one
billion pounds per year in the United States alone, the risks to
agricultural workers and to the general public alike are significant,
especially if one considers that only 1-3% of an agricultural
chemical may actually reach the intended site of action (4). Just
where is the other 97-99% going? Clearly, the application of liquid
sprays to some agricultural crop canopies is a very inefficient
process. Unfortunately, the relationships between the spray process,
the target pests, and the crops themselves are very complex and
certainly not well understood. The same can be said about the

state of the art involving hazardous waste disposal which, of
course, includes pesticide waste management. Pesticide handling,
from inception in the test tube to ultimate disposal of wastes by
applicators is not without risk . . . to the user, to the general
public, and to the environment.
    All pesticide users and handlers need to keep abreast of
current innovations in personal protective equipment, application
technology and disposal techniques.

Personal Protective Equipment

Pesticides as a class of chemicals are toxic. But pesticides need
not be hazardous to the user if ways can be found to reduce
exposure. Whether we are driving an automobile, cutting firewood
with a power saw, or using pesticides, the risks or hazards
associated with these modern technological innovations can be
substantially reduced by effectively utilizing appropriate risk
reduction techniques and practices. One such practice, at least
with regard to pesticides, is to use all appropriate protective
clothing and other safety equipment. Such protective devices can
reduce and, in some instances, eliminate exposure to pesticides
altogether (1). The type of protective clothing and equipment
needed depends on the job being done and the type of chemical being
used. With some of the more hazardous chemicals such equipment is
often mandatory.
    As a minimum the following protective items should be available
when handling pesticides:
    (1) Clean clothing, including a long-sleeved shirt, long
         trousers, and/or coveralls or a spray suit made of a
         tightly woven fabric or a water-repellent material.
    (2) Waterproof gloves, unlined and without a fabric wrist
         band. Shirt sleeves should be worn over gloves in most
         instances, not tucked inside.
    (3) Waterproof boots. Pants legs should be worn over boots,
         not tucked inside.
    (4) Wide-brimmed waterproof hat.
    (5) Goggles or full-faced shield.
    (6) Respirator with a clean cartridge or canister. The
         correct type of cartridge or canister must be used for the
         specific chemical being applied; they differ for
         particular kinds or groups of toxicants.
    Considerable difficulties unfortunately remain with both the
design and use of these protective devices. The independent farmer
who regularly handles and applies pesticides tends to shy away from
using protective garments, often complaining that such clothing is
uncomfortably warm. In hot weather, some types of protective
clothing may actually contribute to heat exhaustion. The same
complaints are often directed against pesticide respiratory
protective devices as well. Unfortunately, the protective equipment
that provides the best protection for the pesticide applicator is
also usually the most uncomfortable and cumbersome to wear, not to
mention frequently the most costly.
    There are now, however, reasons for optimism. Inexpensive
disposable or limited-use lightweight clothing and accessories have
now become commonplace in industry as personal protective items;

i.e. coveralls, coats, shirts, pants, and hoods (5). Spunbonded
olefin (Tyvek) is the most commonly used fabric for industrial
disposable or limited-use protective clothing. Such fabric is made
by spinning continuous strands of very fine, interconnected
polyethylene fibers, and then bonding them together with heat and
pressure (6). The fabric can also be coated or laminated with a
polyethylene coating for added protection. For example, break-
through of a methyl parathion field spray through uncoated Tyvek,
polyethylene-coated Tyvek, and Saranex-laminated Tyvek fabrics was
less than 5 min., 30 to 45 min., and greater then 240 min.,
respectively (7). The coated/laminated spunbonded olefin fabric is
clearly an effective particle barrier and resists liquid splash
penetration, properties absolutely essential when applying
pesticides.
     One of the most desirable attributes of this fabric is its
lightweight property. A typical nonlaminated coverall weighs only
about five ounces, thus eliminating or, at least, reducing the
"fatigue factor" commonly associated with heavy rubberized spray
suits. Because of their superior liquid hold-out properties,
durability, comfort, and low cost, spunbonded olefin and other
nonwoven fabrics are now being used widely in limited use or
disposable protective garments. In the future, with the development
of new fabrics, new combinations, and new style modifications, these
lightweight garments will find even wider usage among pesticide
users as a way to minimize occupational exposure.
     If protective garments can minimize dermal deposition of
chemicals, then respiratory protective devices can aid in reducing
pulmonary exposure to airborne volatiles and particulates. As with
protective garments, the most common excuse for not wearing a
respirator when applying pesticides is comfort. It is generally
easier and certainly much more comfortable not to wear a respirator.
Here again manufacturers must address the comfort factor if
pesticide users are expected to wear even the most basic air-
purifying device, the chemical cartridge respirator.

Application
_____

Whereas protective clothing and equipment attempt to reduce
applicator exposure by intercepting spray droplets or dust particles
prior to deposition on the skin or transfer to the lungs, pesticide
application technology tends to focus on improving penetration and
deposition of agrichemicals into crop canopies or, in other words,
reducing movement off target. New pesticide application innovations
depend on many factors, of which safety is only one. Factors such
as cost, convenience, labor time requirements, and, of course,
efficacy usually take precedence (1). Few types of application
equipment have been tested for their exposure impact, and very few
of the many combinations of formulations and active ingredients have
been tested with each type of equipment.
     During the past 30 years there has been phenomenal progress in
the development of highly active and effective agrichemicals, yet we
have not kept pace in either the development of efficient pesticide
delivery systems or in developing a basic understanding of the
components of efficient pesticide application (8). If development
indeed lacks in these areas, no wonder then that so little research

has been done on how application methods affect pesticide exposure levels.

Although we are well aware that there is an exposure potential from any type of formulation, we generally relegate our major concerns to the liquid or sprayable formulations. When liquid application is essential, the formulation can be modified to assure that droplets remain large enough to minimize drift (1). The worst drift problems result from the smallest, nonvisible particles or droplets, generally those under 100 microns (Table II). The smaller the particle, the greater the drift potential.

Table II. Drift of an Oil/Water Emulsion[1]

Particle Diameter (microns)	Drift Distance Downwind (Ft.)
800	25-50
400	50-100
200	150-300
100	500-1,000
50	Indefinite

[1]2.8 oil/water emulsion applied during a strong inversion at a boom height of five feet and wind speed at 5 MPH.

Spray nozzles designed for both aircraft and ground equipment can also be used to enlarge droplet size of the spray. Application equipment can also be modified to reduce drift. For example, shrouding the spray booms of ground equipment keeps droplets from swirling up into the air, thus reducing the potential for drift and applicator exposure.

A look at two recent technological innovations in spraying systems as well as a current application practice will bring the issue of applicator exposure in relation to application practices and technology into better focus.

Electrostatic sprayers propel charged pesticide droplets to the target crop. In most field tests, significantly improved deposition coupled with a reduction in drift occurred with charged droplets (9). These droplets are propelled at high velocity from the spray nozzle to the target and, because they are positively charged, they are mutually repellent to one another and attracted to the crop. Although more extensive applicator exposure studies are needed using electrostatic sprayers, preliminary results suggest that such equipment improves safety as well (1) (Table III). Significantly improved deposition associated with these sprayers means more chemical is deposited on the target plant (and possibly adjacent soils) with less to drift in the environment and presumably less deposition on the applicator.

Another new method of spray application which may improve foliar deposition is called air-assist spraying (10). This application system involves spraying air along with the pesticide to enhance penetration of crop canopies. With more spray impacting the plant, not only is environmental impact reduced but the performance of insecticides, fungicides, foliar fertilizers, and growth regulators is vastly improved.

The new air-assist system has been shown to improve overall

Table III. Applicator Exposure Using Conventional and Electrostatic Sprayers

Country	Device	Grams active applied per hour	Formulation	Crop height (cm)	TDC* mg/hr	TDC* as % of active ingredient applied
Tanzania	Electrodyn sprayer	16	Cypermethrin	30–60	26.9	0.17
	Spinning disc	116	Cypermethrin	30–60	369.9	0.32
Ivory Coast	Electrodyn sprayer	21.6	Cypermethrin	120–160	8.9	0.04
	Spinning disc	23.4	Cypermethrin	110–180	17.8	0.08
Paraguay	Electrodyn sprayer	6.6	Cypermethrin	66–125	3.0	0.05
	Knapsack	13.1	Cypermethrin	75–180	29.5	0.22

*TDC—Total dermal contamination
Source: T. B. Hart, "The Hand-Held 'Electrodyn' Sprayer: Worker Hazard," ICI Plant Protection Division (Fernhurst, England: undated).

coverage of chemicals on crop foliage by 234 percent on soybeans and 100 percent on corn. The air-assist sprayer uses fan nozzles to produce the spray. About 8 inches beyond the nozzle the spray mixes with swirling air generated by a centrifugal fan to form a mist around the target plant. These accelerated droplets enter the target area of the plant and are deposited where directed. Air-assist spraying can also be done inside a metal shroud, giving good spray saturation plus spray recirculation.

Spray droplets generated by air-assist sprayers are obviously better able to penetrate dense canopies of foliage because of accelerated droplet velocity. But does this added spray momentum also increase topical and inhalation exposure for applicators? The exposure issue with this new technology has not been adequately addressed, yet must be examined before this or any new spray system can be classified as an unqualified success.

A popular current practice is the use of vegetable oils as carriers in the application of crop pesticides. As long as the product label specifically includes instructions that permit the use of vegetable oils, such as cottonseed and soybean oils, as diluents or carriers in place of water, it is perfectly acceptable to use the pesticide in this manner. This applies to diluents used in conventional, low volume, and ultra-low volume applications. As this practice rapidly increased, EPA expressed concerns about the practice relevant to applicator/farm worker safety (11). The use of vegetable oils as spray diluents/carriers might result in chemical residues on crops in excess of permissible amounts as well as increase farmworker exposure to the pesticides when handling treated crops. EPA officials noted that vegetable oils evaporate more slowly than water, and there are longer-lasting residues after the materials have been applied. The oil may also increase the human body's absorption rate of the active ingredient, further prompting worker safety concerns.

In response to these concerns, the Office of Pesticides and Toxic Substances, EPA, issued FIFRA Compliance Monitoring Policy No. 12.5 on February 27, 1984 (11), which in summary states that in such instances where no diluent is specified on the label, water must be used as the diluent. Clearly, this response was prompted by health and safety concerns for pesticide applicators and farmworkers.

Pesticide Waste Disposal

The issue of pesticide waste disposal has been recognized as a national problem for years, yet today remains as one of the foremost problems confronting most pesticide users. In any Extension meeting that addresses pesticide safety issues, the most frequently discussed topic is that of pesticide waste disposal. The undisputable fact is that adequate hazardous waste disposal facilities do not presently exist for small volume pesticide users. Improper, albeit not necessarily irresponsible, handling of pesticide wastes and containers often results in unacceptable levels of environmental contamination and excessive exposure to the applicators themselves.

Pesticide waste disposal policy and practices have been dealt with recently by three national workshops (12, 13, 14). These conferences defined the problem and examined the state-of-the-art

technology for pesticide waste disposal. The sources of potential
problems include the containers; unwanted, unuseable and
unidentifiable products; tank rinse waters; leftover materials;
equipment wash waters; incompatible mixtures; spilled materials from
accidents; stormwater and run-off from natural occurrences; and
toxic debris from fires (15). Defining the problem is relatively
easy; it's the solutions that are so difficult to develop.
     Speakers at these national workshops described container
collection programs and various techniques for waste treatment and
disposal. Much of the discussion was devoted to pioneering efforts
to deal with disposal problems. For the most part, however, the
issues of environmental and human health concerns were not
addressed.
     Of the dozen technologies that are being investigated for
disposing of pesticide wastewater, only four methods are currently
available commercially (13, 16) (Table IV). But, the question must
be asked, available to whom? Often even the simpliest technology is
not priced in the range that farmers and small commercial
applicators can afford.
     Recycling pesticide rinsewater seems to be a wastewater
management system that could be available to most pesticide users.
The basic design of a wastewater collection/recycling system
includes a wash pad and some type of receptacle for rinsewater
containment. The rinsewaters can then be mixed as needed into
subsequent spray solutions as an alternative to treatment or
disposal. However, this practice does elicit some concerns
especially where more diverse and sensitive specialty crops are
grown. The main concerns arise from tank mixing of pesticides not
labeled for such use, phytotoxicity to some crops, and residues in
excess of established tolerances (17). The potential for excessive
applicator exposure must also be viewed as a significant concern
because the rinsewaters, laden with these chemicals, are used much
the same as tap water is used for mixing chemicals in a spray tank.
This presents an additional exposure risk to the applicator every
time the rinsewater is handled during the recycling process.
     Closed system technology for transferring pesticide
concentrates into a spray tank is designed to empty shipping
containers of liquid pesticides, rinse these containers, and
transfer both product and rinsate to the application equipment while
effectively reducing applicator exposure by minimizing spilling and
splashing of pesticide concentrates during mixing and application.
Possibly a modified type of "closed system" would be appropriate for
storing and ultimately recycling pesticide-contaminated rinsewater
to the spray tank. Such handling and storage techniques would
certainly reduce the potential for applicator exposure.
     Aboveground soil disposal beds may ultimately provide an
environmentally safe and economical means of disposing of pesticide
contaminated rinsewater (18). Rinsewater would be collected
initially in a sump and pumped into a specially designed tank with a
lower liquid storage container and an upper layer of soil suspended
on a perforated platform. A sump pump would be used to apply daily
doses of the accumulated liquid to the soil surface via a surface
distribution system. Although it remains to be seen if these above-
ground soil digestion systems will be practically functional, it is
interesting to observe that such a system that includes wastewater

Table IV. Categorization of Pesticide Wastewater Disposal
           Technologies

Technology	Proven Technology[1]	Technology Transfer[2]	Emerging Technology[3]
Physical/Chemical Treatment & Recycling			
1. Pesticide Rinsewater Recycling	X		
2. Granular Carbon Adsorption	X		
3. UV–Ozonation		X	
4. Small–Scale Incineration			X
5. Solar Photo–Decomposition			X
6. Chemical Degradation		X	
Biological Treatment & Land Application			
1. Evaporation, Photo–degradation & Biodegradation in Containment Devices	X		
2. Genetically Engineered Products			X
3. Leach Fields	X		
4. Acid & Alkaline Trickling Filter Systems			X
5. Organic Matrix Adsorption & Microbial Degradation			X
6. Evaporation & Biological Treatment with Wicks			X

[1] Technology is currently being utilized on a commercial basis to treat and dispose of dilute pesticide wastewater.

[2] Technology is being utilized commercially to treat other types of waste and offers promising opportunities for pesticide wastewater.

[3] Technology is not being utilized commercially but experimental data indicates it is a promising candidate technology for pesticide wastewater.

collection and storage capabilities as well as automatic transfer of
the liquid waste to a digester would also reduce direct applicator
exposure significantly.

At the January 1985 workshop in Denver, O.R. Ehart said,
"Although this conference has dwelt upon the regulation of pesticide
waste disposal, it is shortsighted not to recognize that the purpose
of these regulations is not, or at least should not be, to regulate
pesticide waste per se but to protect the environment" (19). The
term environment as used in this sense most certainly also includes
persons who are subject to pesticide contamination through
occupational exposure as well as persons who are exposed
inadvertently through incidental exposure such as might occur
through contact with contaminated potable water sources.

Affordable, innovative technology to remedy the pesticide
disposal dilemma will, with time, be available to the farmer and
other pesticide users. However, as with any new information or
technology, the focus will have to be on education if such technical
innovations are to be used in an efficacious and responsible manner.
Any program on pesticide waste management must also include
information on minimizing the risks to the applicator as well as to
the environment. Nothing less will be acceptable.

Summary

EPA's guidelines do not presently address the issue of measuring
direct exposure to pesticide applicators (20). Rather, the Agency
often takes the "surrogate chemical approach," which uses data from
one compound to set exposure standards for another. The problem
arising from this kind of approach is that such procedures often
fail to provide us information about the permeability properties of
the compounds. The EPA needs to develop a standard methodology for
assessing applicator exposure and issue relevant guidelines to
address the problem.

Pesticide manufacturers and user industries need to place
greater emphasis on health surveillance programs for their
employees. Research programs need to be expanded and coordinated on
assessing exposure to pesticides. While the work on exposure should
be conducted by pesticide manufacturers, university-based
agricultural scientists, and research-oriented governmental
agencies, it still remains the responsibility of those agencies concerned with occupational safety and health to take the overall
responsibility for establishing research guidelines and coordinating
research objectives. However, to accomplish these objectives, the
overall research climate and support for such programs need to
improve rather dramatically in the immediate future.

Literature Cited

1. Dover, M. J. "A Better Mousetrap: Improving Pest Management
    for Agriculture"; World Resources Institute: Washington, D. C.,
    1985; p. 84.
2. Morgan, D. P. In "Residue Reviews"; Gunther, F. A., Ed,;
    Springer-Verlag: New York, 1980; Vol. 75, pp. 97-102.
3. Davies, J. E. In "Determination and Assessment of Pesticide
    Exposure"; Siewierski, M., Ed.; Studies in Environmental
    Science No. 24, Elsevier: New York, 1984; pp. 67-77.

4. Hall, F. R.; et. al. In "Improving Agrochemical and Fertilizer
    Application Technology"; Hall, F. R., Ed.; Agricultural
    Research Institute: Bethesda, MD, 1985; pp. 15-23.
5. Goldstein, L. Natl. Safety News 1980, 50-1.
6. "The Properties and Processing of TYVEK Spunbonded Olefin,"
    DuPont Tech. Inf. Bull. S-10, 1973.
7. Schwope, A. D. "The Effectiveness of TYVEK and TYVEK
    Composites as Barriers to Methyl Parathion," Arthur D. Little,
    Inc., 1980.
8. Bukovac, M. J. In "Improving Agrochemical and Fertilizer
    Application Technology"; Hall, F. R., Ed.; Agricultural
    Research Institute: Bethesda, MD, 1985; pp. 25-38.
9. Matthews, G. A. In "Improving Agrochemical and Fertilizer
    Application Tehnology"; Hall, F. R., Ed.; Agricultural Research
    Institute: Bethesda, MD, 1985; pp. 39-52.
10. Richardson, L. Agrichemical Age 1985, 29(9), 8-9, 12.
11. "The Use of a Diluent Not Specified on the Product Label,"
    FIFRA Compliance Monitoring Policy No. 12.5, U.S. EPA, 1984.
12. National Workshop on Pesticide Waste Disposal: Denver, CO; Jan.
    28-29, 1985.
13. Pesticide Wastewater Research Workshop: Cincinnati, OH; July
    30-31, 1985.
14. National Workshop on Pesticide Waste Disposal: Denver, CO; Jan.
    27-29, 1986.
15. Ehart, O. R. Proc. Natl. Workshop on Pesticide Waste Disposal,
    1985, pp. 2-11.
16. Bridges, J. S.; C. R. Dempsey. "Proceedings: Research Workshop
    on the Treatment/Disposal of Pesticide Wastewater"; U.S. EPA:
    Cincinnati, OH, Jan. 1986; p. 55.
17. Taylor, A. G. Abstr., Pesticide Wastewater Research Workshop:
    Cincinnati, OH; July 30-31, 1985.
18. Brown, K. W. Agrichemical Age 1986, 30(1), 14, 44.
19. Ehart, O. R. Proc. Natl. Workshop on Pesticide Waste Disposal,
    1985, pp. 120-3.
20. Wasserstrom, R. F.; Wiles, R. "Field Duty: U.S. Farmworkers
    and Pesticide Safety"; World Resources Institute: Washington,
    D. C., 1985; p. 78.

RECEIVED September 19, 1986

Chapter 12

Educating the Public Concerning Risks Associated with Toxic Substances

Ronald W. Hart and Angelo Turturro

National Center for Toxicological Research, Jefferson, AR 72079

A comprehensive program modeled after the present Department of Health and Human Services (DHHS) recommendations for a problem-solving approach in risk assessment/risk management would be useful in educating the public about the risks from exposures to certain chemicals, e.g., pesticides. Documents on the nature of risk and risk perception, as suggested in the DHHS recommendations, can be useful in putting risk into perspective. Suggested is a "problem-solving approach", which not only separates the evolution of a risk management decision into an analysis of the risk assessment, but also clearly states the uncertainties in the risk assessments. This "problem-solving" approach also attempts to provide comparison and justification to other governmental risk assessments and the analysis of possible options, e.g., government regulation, risk reduction by technologic means, etc. Additionally, in order to make clear the nature of the risk, it offers a plan for informing the affected parties, other health-related agencies, e.g., federal, state, and local health agencies, primary care physicians, and suggests a plan for the evaluation of the option selected. It is suggested that these issues be explicated in special sections of a risk management document, and in language aimed at the layman. It is strongly suggested that the risk assessment/risk management document be used as an instructional tool to assist both the public and risk manager in deciding how to evaluate the significance of the risk.

The ancient Babylonian lived in a world populated by terrifying spirits, a world fraught with perils derived from the slighting of jealous gods. In the villages, mystical shamen/prophets interpreted the will of the gods in capricious and whimsical ways based on bizarre extrapolations from personal observations. In such an atmosphere, the Asipu, the first risk assessors known to us (1), delivered pronouncements on proper planting times, risky ventures, etc., based on a complex system using pluses and minuses derived from divination results. The Asipu thought they divined the will of the gods, and their authority, bulwarked by the regency, convinced the public to comply with the results of their assessments. The authority of the king and priests was important as a counterweight to a frightening cosmos.

Although this example is from the beginning of recorded history, aspects of it are uncannily familiar. The terrifying spirits now are toxic agents (the "sea" of carcinogens) (2). The mystics in the public marketplace today speak of idiosyncratic interpretations of test results (e.g., human danger from the marginal observation of chromosomal changes in fungi) or of personal experiences ("My cow stopped supplying milk when she was exposed"). The assessors, speaking with the authority of government, now use computers, and quantitate economic dislocation as well as physical destruction. Similarity to both the ancient and modern situation is that unknowable phenomena are being interpreted through ways approaching magic to the general public.

A fundamental break with this tradition has been recommended recently by the Department of Health and Human Services (DHHS) "Report on Risk Assessment and Risk Management in DHHS" (3). Instead of requesting that the public "trust us", or requiring a personal mystical union with some embodiment of "Nature", this document recommends that the public be brought in as a partner to understand the meaning of risk, the method for assessing risk, and the response to risk. This "democratization" was not to occur by mindless extrapolation of a single test result, such as cancer induction in a single test, but by promoting a true understanding, as much as possible, of the state-of-the-science of a complex process. This is to be done by a "problem-solving approach" to risk assessment and management, which seeks, to educate the public, as much as possible, concerning the risks from chemical exposure in order to achieve an effective risk management strategy.

Although originated for DHHS, the approach has potential for use throughout government when it is necessary to communicate risks to an interested public. This is especially true for pesticides. These chemicals, which are an important part of the armamentarium against that aspect of nature which would take the very bread from our mouths, have acquired a somewhat unsavory reputation because of the potential risk associated with some of their chronic effects. The method chosen to explain the context of a pesticide risk (health hazards, etc.) to the public is a vital part of effective pesticide use management.

A tactic in the educative process is to use risk assessment and risk management documents as teaching tools to train the public to understand and manage risk, as noted below.

Definitions

There is no surer block to communication that using "common" terms which have different meanings for different individuals. This is especially true in risk assessment, especially for different

government agencies. For instance, if one agency is discussing
risk in terms of adverse human health effects, while another is
addressing the issue in terms of the effects in animal studies,
this can lead to one agency saying there is no risk, while another
declares an emergency. One of efforts in the DHHS document (3) was
to develop a common set of definitions for the different agencies
in DHHS. These definitions are adapted from the NAS report on risk
assessment (4) and the OSTP Report on chemical carcinogenesis (5).
     Risk - The probability of an adverse health effect as a result
of exposure to a hazardous substance(s).
     Risk Assessment - The use of available information to evaluate
and estimate exposure to a substance(s) and its consequent adverse
health effects. Risk assessment consists of one or more of the
following four elements:
     Hazard Identification - The qualitative evaluation of the
adverse health effects of a substance(s) in animals or in humans.
     Exposure Assessment - The evaluation of the types (routes and
media), magnitudes, time and duration of actual or anticipated
exposures and of doses, when known; and, when appropriate, the
number of persons who are likely to be exposed.
     Dose-response Assessment - The process of estimating the
relationship between the dose of a substance(s) and the incidence
of adverse health effects.
     Risk Characterization - The process of estimating the probable
incidence of an adverse health effect to humans under various
conditions of exposure, including a description of the uncertainties involved.
     Risk Management - The process of integrating risk-assessment
results with engineering data and social, economic and political
concerns, then weighing the alternatives to select the most
appropriate public health action, ranging from public education to
interdiction, that will lead to reduction or elimination of the
identified risk.
     In the DHHS report, it was suggested that these definitions be
the ones generally used when communicating the risk from chemical
materials to the public. Some general consensus definitions,
agreed to across government can be very useful in preventing
inconsistency when presenting the information on hazard from a
chemical, e.g., a pesticide, to the public.

Factors in Understanding Risk

To understand the nature of the societal response to risk, it is
important to appreciate a number of factors which influence this
response.

Acceptability of Risk. Although risk has always been part of life,
the acceptability of different kinds of risk has varied considerably. For instance, deadly outbreaks of infectious diseases were
once a sign of God's displeasure and, therefore, acceptable.
However, with the advent of modern public sanitation, vaccines, and
other modern technologies, very limited outbreaks, such as that
experienced with Legionnaire's Disease, which would have hardly
been noticed in the recent past, are now considered national emer-

gencies. Mining, once done by "disposable" slaves because of the attendant hazards, has become an occupation in which substandard working and health conditions are not tolerated. Much of this change is the fruit of the efforts of countless physicians, politicians, labor leaders, regulators and journalists. As a result of efforts such as these, and the improvement in living standards, the major focus now is on non-infectious diseases, such as most cancers, and the health consequences of the industrial processes from which this improved standard of living is derived, such as the consequences of pesticide use. An important factor in the public acceptibility of a risk is also the confidence in the societal ability to control it (6). Risks which are not readily controllable strike much more fear than those that are. As exposures alter, as mores change, as prevention and control techniques improve, as laws evolve, as needs change, as alternatives become available, as information on hazards improves, so then the acceptability of a risk changes.

Estimation of Risk. The estimation of risk contains uncertainties, based on the lack of specific data (such as exposure information) and/or the lack of understanding of the mechanism of toxic action of a compound. Between the extremes of acturial risk, which is based on enough information that "time has removed the uncertainty," such as the probability of death as cited in an insurance table, and theoretical risk, which is based on probabilistic calculations of events which have never actually occurred (e.g., nuclear "winter" (7)) lies a wide continuum into which most estimates of human health effects fall. In real-life situations, many assumptions are made in evaluating risk in order to make a conclusion, and these assumptions lead to uncertainties in the final result. These uncertainties should be understood as limitations to the best guess science can presently make. Although one response to this uncertainty, in the face of an outcome as fearsome as cancer, is to deny that there is a lack of certainty, the more reasonable response is to try to estimate the uncertainty, making it clear that any estimate is bracketed by these possible errors.

    Also important to consider is the nature of the risk. When a physician recommends that a patient change his lifestyle to reduce risk of heart attack, the physician is synthesizing clinically-oriented human research and personal experience with an expert evaluation of the relevant personal characteristics which bear upon the hazard. This personalized risk management, based on what can be termed "personal risk," is very different than the approach based on population risks contained in risk assessments. The population risks are almost always upper confidence bounds, often based on "worst-case" scenarios. Many important characteristics in assigning risk have not been identified, and personal qualifiers are ignored. Frequently, there is little, if any, epidemiological data to demonstrate human effects. A source of confusion is to consider risks of all types as immediate and personal risks. Personalizing inferential population risks can lead to the conclusion that everything is dangerous, that one is treading in a veritable minefield where the slightest misstep can explode into a horrible cancer. This perception does not accurately see population risks simply as part of the process of prioritizing and managing risks on a broad scale.

Voluntary Aspects. People voluntarily accept some risks, such as driving a car, and have others imposed upon them, such as water pollution. The line between voluntary and involuntary risks is often hard to define and is frequently determined by availabilty of resources and social practices, e.g., if one has a filter, one could drink only filtered water. In general, the American public, probably reflecting our individualistic biases, is more tolerant of voluntary risks (8). Americans do not readily accept their government to be the arbiter of personal risk, demonstrated by the brouhaha that developed over governmental attempts to mandate use of seat belts and motorcycle helmets (9). (For a good discussion of the broader concept of risk and consent, see 10). Much more acceptable appears to be government efforts to protect the public from imposed risks.

Perception of Risk. Fairly recently, it has been appreciated that public perception of risk is important in risk policy (11). Some of the major conclusions that can be drawn about public perception are (12,13):
a) Cognitive limitations, coupled with anxieties generated by the feeling that one is gambling with one's life, cause uncertainty to be denied, risks to be distorted, and statements of "fact" to be believed with unwarranted confidence.
b) Perceived risk is influenced by the imaginability and memorability of the hazard. In this aspect the media has a special role since it can make the unimagined real, vivid, and fearsome. For instance, publishing a series of articles on birth defects, with pictures of deformed babies, is likely to heighten the sensitivity of a community to information about contaminants in water that may be teratogenic.
c) While safety experts tended to perceive risk in a manner closely responding to the statistical frequencies of death, lay persons' risk perception included aspects such as dread, the likelihood of fatality, and the degree of catastrophic potential. For instance, the public perception of the risk of death by flood is high compared to the danger of asthma, a much more significant killer (14). This difference in perspective is especially evident for chemicals. For instance, a synthetic chemical such as trichloroethylene (TCE) is treated as a very dangerous entity, while aflatoxin contamination of food or the effects of drinking alcohol, both with orders of magnitude greater cancer risk than TCE (15), is considered relatively benign.
Public perception of risk, therefore, can vary significantly from that of safety experts. This difference in perception is important to evaluate in a risk management strategy.

Understanding Risk. It is important that the public understand the nature of risk. Explaining and illustrating the risks involved in toxic exposure, plus relating them to the risks of everyday life, is crucial if the public is to understand how to put risks for toxic exposures in context. Context is vital if people are to get an accurate picture of what threat a hazard presents and to whom. It is useless to note that the upper limit on the risk associated with a particular substance is on the order of 10   unless one also gives the appropriate context for such a number. The population at elevated risk may be miniscule, however, the risk could be misinterpreted as a general population danger. A risk assessment/risk management document should do this, and provide the scientific bases for the explanation by risk managers of the risk to the public. This information, placed in an organized fashion in the document, provides the bases for those who interpret the meaning to the public directly, such as private physicians, health departments, company health managers, etc. The object is not to turn the public into professional risk assessors and/or managers, but, understanding the many-faceted interests in the many individuals who comprise the public, is to encourage active participation by interested parties in managing risk and participation by as many members of the public as possible.

Problem-solving Approach to Risk Management

The major challenge in risk management is to enhance public welfare through effectively managing the risk of a chemical that has toxic effects under practical conditions of use and exposure, i.e., a successful risk management decision. In order to manage risk effectively, one must have an adequate assessment of the situation, and realistic plans for coping with the hazards that derive from the situation. A crucial element for success is public understanding and cooperation in all aspects. A systematic process separates the analysis of a risk assessment/management decision into four sections:

1) analysis of the risk assessment;
2) analysis of possible options;
3) promotion of understanding and acceptance of risk-management decisions; and
4) evaluation of the effectiveness of the options chosen.

and asks whether each section completes its task in terms of leading to a effective risk managment strategy.

This process can be termed a problem-solving approach to risk management. Some of the formal rubrics termed "problem-solving" tend to be quite general, e.g., using such concepts as problem-identification, determining important parameters, etc. Problem-solving referred to here is quite specific to risk assessment/risk management as presently performed and focuses on its goal, a successful risk management decision, with the appreciation that public understanding is a key portion of an effective strategy. It is an approach that attacks each risk management decision as a "clinical research experiment" in resolving toxic-related situations in a manner most conducive to public welfare. The scientific method is incorporated as much as possible and the approach "learns" in order to expedite the next decision. Aspects of this approach are currently being used in different contexts by different units in government, and this systemization seeks to place these efforts in context.

Analysis of the Risk Assessment.   In answering the question of
whether the risk assessment is adequate for basing a risk
management decision, a number of issues arise.   Some are listed
below.

1) A risk management decision should be based on the clear
understanding of the limitations of the risk assessment.   This is
difficult, if not impossible, if the risk assessment doesn't
characterize the uncertainties comprehensively, e.g., identify the
result of altering different assumptions which are bases of conten-
tion; the major sources of uncertainty, etc.   This could become a
special section of the risk assessment document and would be
invaluable to risk managers and the public in understanding the
validity of the assessment.

2) In order to evaluate options, it is important to understand what could be done to limit the uncertainty, especially when
this radically changes the options.   For example, if the exposure
level for safety is based on a conservative approach, necessitated
by the uncertainty in an assessment, and an experiment could reduce
the uncertainty by a factor of ten, then the control level may be
able to be set much lower.   This could change the control options
significantly.   A level of 10 ppb could require destruction of 1/2
of a crop, while a level of 100 ppb could have little, if any,
economic consequences.   For the present data deficiencies which
contribute to the major uncertainties in the assessment, it is
useful to identify the research, if any, which could limit uncertainty.   Research such as this could lead to re-evaluation of a
risk assessment after data are obtained, making delay of a decision
unnecessary, and leading to increased use of relevant information
in a risk assessment.

In the special category of contributing to public understanding of the risk management decision, understanding of the risk
assessment would be heightened by:

1) The bases of the risk assessments, i.e., the assumptions
which underlie the process, should be elucidated in plaintext
(i.e., simple, straightforward common language) as much as possible.
A special section in the risk assessment document should be written
to comprehensively discuss the assumptions, and could also be in
plaintext.

2) Understanding risk management decisions in light of major
differences in risk assessments by various agencies is particularly
difficult when the reasons for the differences are not clearly
presented, and can be a major stumbling block to public cooperation.   It is crucial for public understanding that the assessment
be compared with other risk assessments for the same compound.
Plaintext explanations of the reasons for any differences (probably
a result of different assumptions being used) will be very
important in the education process.

Analysis of Possible Options.   The next step is to question whether
the risk management decision actually considers all possible
options and chooses one which maximizes public welfare and effectiveness.   The decision is seen as an opportunity for government
expertise to protect public health with minimal losses (better if
everybody gains).   Analysis should not be limited to economic and

social benefits, but should also include health benefits (e.g., it would make little sense to transport a material across the country if the public health dangers of the transfer exceeded those of letting the material remain where it was).

Some examples of the range of options to be considered are:

1) Agency regulatory action - which can range from public education, warning labels, etc. to interdiction and total ban, with plaintext explanations of why the particular route was chosen;

2) Regulation by other governmental units - Evaluation of the role that state, local and other federal agencies play should be clearly set forth. This leads to coordinated action, with removal of any contradictory regulations and different units working at cross-purposes. Discussing how different control possibilities would imapct on different agencies is an important part of this evaluation.

3) Risk reduction by technologic means - Technologic strategies to reduce risks should be considered, e.g., detoxification, filtration, alternatives, etc. Personal protective gear, changes in application procedures, changes in formulations are technological solutions especially relevant to minimizing pesticide risk. However, when discussing technologic solutions, it is useful to discuss them in context of the new risks. For instance, a new application using different ingredients may simply exchange chemicals with no relevant information for those for which we have limited information that they are toxic.

4) Voluntary actions by the private sector - "Jawboning", i.e., discussion of the problem with concerned interests, e.g., the manufacturer, applicator, etc., can be very effective in maximizing compliance and speeding up the removal of any health hazard. This approach also maintains maximum flexibility as the "regulations" are not fixed.

5) Actions by professional societies - A long history of public service characterizes the professional societies. This avenue should be formally considered when faced with the need to resolve especially complex issues. These societies are a valuable resource in reviewing documents, assisting in clarifying consensus positions, etc. and as a source of ideas for coping with difficult situations.

6) Risk management research - Similar to the need to identifiy research in the risk assessment, identifying risk management research involving the social sciences, research into the factors which contribute to the success of various options, especially in public health terms is important. It has been relatively neglected. However, when risk management decisions are considered attempts to resolve situations, an obvious question is how to maximize application of management strategies.

Again, in the special case of increasing public understanding, it is important to demonstrate that a wide range of options was considered, and those with the least risk and most benefit chosen. It is useful for acceptance to clearly state the reasons for choosing a particular option, both its advantages and disadvantages, especially in the health area, as the different segments in the society will seek out the reason for a decision which will have impact on it. Very useful would be a plaintext statement of the

nature of the risks, benefits, options, and factors (e.g.,
statutory mandates). A special section in plaintext would be very
useful for this.

Promotion of Acceptance of Risk-Management Decisions.   In a risk
management decision, it is important to ask how to maximize
acceptance of and compliance with a risk management decision,
although this is often overlooked. Too often it is forgotten that
a reasonable explanation can sometimes be more effective than
police action. Examples of actions are below.

1) Getting the affected parties to participate in and understand the major aspects of the decision as soon as possible. This
is different than general public understanding since the response
to the issues will probably be much more volatile and heated.
Crisis teams especially trained to explain and adapt decisions,
when possible, are probably the best way to accomplish this. What
is needed is not spokesmen, but active participants who can modify
and adapt the risk management.

2) Getting information to local units.   Dissemination of
information to health-related agencies in the area, as well as
primary care physicians on who is likely to be affected and in what
way, is important. These individuals will probably be the front
line addressing many of the questions involving health, and need
all the health information the federal government can provide.
Working with these individuals as partners, as soon as possible
(perhaps when options are being considered) can mitigate problems
beforehand.

Evaluation of the Effectiveness of Options Chosen.   When considering a risk management decision, it is important that the question
of how to evaluate the decision is addressed. Information on the
efficacy of the option chosen is important both to the process of
choosing future options and to evaluating the need for corrective
action. Factors to evaluate:

1) Effect on health should be evaluated. It would also be
useful to evaluate the cost-effectiveness of the option. A follow-
up provides a record for any corrective options. Further, if
conditions change or the follow-up shows the resolution is
ineffective, the risk manager may decide to alter the decision.
This is an important part of the problem-solving approach which
would lead to improvement. In this sense, the process "learns."

2) For increased public understanding, it is important to
show how effective an option was to improve future compliance. It
also can lead to improvement of the solution through direct public
input and participation.

Conclusion

The role of the public in risk assessment and risk management in a
democracy such as ours is as a partner. Government has an obliga-
tion to explain its actions in a manner as conducive to public
understanding as possible. Part of the societal investment in any
risk assessment/risk management decision should be dedicated to
explanation to those who are to live with the decision. In terms

of efforts to derive experimental results to use in risk assessment/management, broad-based support across the regulatory agencies should be primarily for programs for research to limit the uncertainty in the critical assumptions used in risk assessment and for basic information in risk management (such as evaluation of risk management options). These routes are the best avenues to provide the information necessary to assess and manage risk effectively, and allow cogent presentations to the public. Approaches which attempt to maximize public understanding and participation, coupled with a strong research program to confine many of the uncertainties in the complex process of risk assessment/management, are necessary to truly accomplish the objective of an effective risk management strategy.

Literature Cited

1. Covello, V. T.; Mumpower, J. Risk Analysis 1985, 5, 103-120.
2. Efron, E. "The Apocalyptics- The Big Cancer Lie"; Simon and Schuster, 1985.
3. "Risk Assessment and Risk Management of Toxic Substances A Report to the Secretary Department of Health and Human Services," U.S. Department of Health and Human Services, April 1985.
4. "Risk Assessment in the Federal Government: Managing the Process," National Academy of Sciences, 1983.
5. Office of Science and Technology Policy. Fed. Reg. 1985, 50, 10372-442.
6. Starr, C. Risk Analysis 1985, 5, 97-102.
7. Carrier, G.F. Issues in Science and Technology 1985, 1, 114-117.
8. Thompson, S. C. Psychol. Bullet. 1981, 90, 89-101.
9. Robertson, L. S. J. Communication 1976, 26, 41-45.
10. MacLean, D. Risk Analysis 1985, 2, 59-67.
11. Covello, V. T.; Menkes, J.; Nehnevajsa, J. Risk Analysis 1982, 2, 53-58.
12. Slovic P.; Fischoff, B.; Lichtenstein, S. Risk Analysis 1982, 2, 83-93.
13. Slovic P.; Fischoff, B.; Lichtenstein, S. Environment 1979, 21, 14-39.
14. Lichenstein, S., Slovic P., Fischhoff, B., Layman, M., Combs, B. J. Exp. Psychol.: Human Learn. Mem. 1978, 4, 551-563.
15. Ames, B.N. Science, in press.

RECEIVED September 16, 1986

Chapter 13

Mass Media's Effect on Public Perceptions of Pesticide Risk: Understanding Media and Improving Science Sources

JoAnn Myer Valenti

University of Tampa, Tampa, FL 33606

Mass media rarely change existing attitudes, but
through reinforcing messages and repetition they
do strengthen public perceptions thereby impacting
behavior. In assessing the role of media in public
perceptions of pesticide risk, this presentation
examines the process of mass communication and the
relationship between journalists and the science
community. Accuracy, objectivity and sourcing are
identified as problems. The author offers as solutions an understanding of the media process and the
constraints faced in covering science stories, as
well as, an improved performance by scientists as
sources of information. Major questions considered
include how media define news, who media use as
sources, and whether media can be expected to
cover risk. The presentation relies on available
studies and analyzes recent cases of media attention to pesticides. Print and broadcast media are
discussed.

At a recent gathering of journalists and representatives from science
and industry, one biologist snapped from the audience, "I don't understand your business and I don't expect you to understand mine!"
That's the sort of useless attitude guaranteed to continue the
current cold war that exists between mass media and science. Just
as journalists are training themselves to better understand the process of scientific thinking, science professionals ought to be working at understanding how mass media function.
  Reporters are consciously struggling with the liabilities of not
enough time and inadequate knowledge to cover science stories.
And they are doing well according to the findings of the Twentieth
Century Fund's 1984 Science in the Streets study, and more recently,
the Scientists' Institute for Public Information (SIPI) survey. In
SIPI's 1986 study, over 90 percent of the scientists queried felt
that journalists were performing with accuracy and appropriateness,
and were helpful in improving the public's understanding of science.

So what is the problem with media's science coverage, and in particular, are there problems with media's coverage of pesticides or the
risk associated with pesticide use? Stuart Diamond of The New York
Times reasons, "Zero risk does not exist...risk is basic to technology." This reporter sees his task as one of balancing risk with
benefits and resisting the short hand explanation, the convenient
line that may not be accurate in each case. But in the effort to
find the answers and detail, what many journalists confront is experts who are afraid of the press, and scientists who don't understand media process or pressures. In reality, the information
vacuum gets filled, even if it comes from non-experts who are willing to talk to the media rather than stonewalling with a "no comment."
     Government or industry employed chemists as well as scientists
in general are considered "a hostile audience" by media. It is
believed that scientists have no respect for reporters. "There's
no learning in the face of all of that hostility." But since I
decided some time ago that I'd rather work with those who are attempting to act as sources for media--trying to access media--rather
than singularly side with journalists, I took this assignment. This
symposium and the long term efforts of the American Chemical Society
such as the ACS news service make it clear that many scientists as
sources do want to improve their relationship with media representatives and improve their ability to use mass media to effect the public perception of pesticide risk.

BACKGROUND

Journalists are not comfortable with the what-if story. There's
greater safety and potential for objectivity in reporting what has
already happened. Uncertainty in a response from an expert usually
does not get into the story. Even worse is the scientists' reluc-
tance to speculate. Journalists are beginning to recognize that the
potential risk question probably should not be aimed at the scientist. But at whom then? Who will speculate in regard to risk for
the next morning's paper? What you do wrong as sources hurts. What
you do right as sources will make the difference.
     Reporters know that readers probably will not understand the
water and phosgene and methyl isocyanate reactions covered in the
Bhopal story. The story gets summed up as an unnatural disaster and
pesticides become only a party to the larger issues of worker training, safety regulations, corporate responsibility, and so on.
Whether the product is pesticides, nuclear power or poultry becomes
secondary, although the association can become enormously detrimental. However, blaming coverage of a tragedy for the actual event
is absurd. In the Bhopal story, the press managed to obtain the all
important company manual that defined methyl isocyanate (MIC) and
provided the instructions for its safe use and production. That
manual was essential to the coverage of the story. The residents of
Bhopal on the other hand reported to journalists that they thought
the Union Carbide plant was producing "medicines for the crops" and
had never been given brochures about MIC or provided with educational
or even public relations literature before the accident. Effective
mass communication means more than answering press questions after
something has happened.

A journalist's human sense perception is operative in covering any story. What looks are on the faces of those involved? How do things in general look? How does the air smell? Some of the best reporting comes in the form of analogies. Diamond wrote in covering Bhopal: "To make the pesticide (Sevin) carbon tetrachloride is mixed with methyl isocyanate and alpha-naphthol, a coffee-colored powder that smells like mothballs." Where did he get that description? Did the sources he was interviewing treat him as a human observer and offer some assistance in explaining complex chemicals and procedures to his readers, or is Diamond on his own expected to be a good science writer?

Jim Detjen of the Philadelphia Inquirer acknowledges the hard feelings and difficulty in confronting these complex stories when he says, "No one wants to sound foolish." Experts are ego-involved in a professional ethic that does not prepare them for quick, simple responses. So, "they clam up." Oftentimes an expert really isn't sure and doesn't want to say, "I don't know." There's a sense that reporters have fried too many experts with a printed, "No comment."

Editors and reporters want to get the story right. There were changes in the New York Times Bhopal series, even changes in the lead paragraph, up to the very last minute--at one point a change in the lead between editions of the paper. Journalists are as committed to accuracy as a science researcher. The overwhelming difference is the immediacy of the journalists' need for answers.

"We're covering issues that are suddenly thrust into the public domain; we dig and find untruths, we still go for a balanced story and then we get attacked by the chemical industry and the American Chemical Society," one reporter told me. It comes as no surprise that respect is lacking in the relationship from both sides. What needs to be confronted here is that the industry is ultimately responsible for its own image. The messenger can't be shot for the problems carried in the mailbag.

THE PROBLEM

The American Chemical Society is not the first to look for flaws in the media. The American print and electronic media catch the blame from computer manufacturers for the furor over a faulty chip, from bankers for contributing to bankruptcy rumors, from the nuclear industry for the failure of nuclear power, and from the government for everything that's wrong. Yet studies continue to show that the public have faith in the media and research indicates that media coverage is in fact fair and unbiased. This paper attempts to offer a very basic understanding of media's approach to coverage of pesticide risk and how the American Chemical Society can work with media to improve that coverage.

Ronald Kuhr, an editor of this book and the head of the Department of Entomology at North Carolina State University told me, "I myself don't like to talk to reporters. It destroys your credibility as a scientist." Kuhr echoes a common complaint when he accuses journalists' translations of what was said of being wrong. He believes, as do some other scientists, that while scientists tend to be very precise, journalists are not precise, and their translations change meanings. Before the translation, it is important to understand the reporter's involvement in the first place.

Where does news begin? What are the news pegs? (What are the significant events or what is it that's new?) The answer usually is that news begins with an accident. (Fires are usually a guaranteed news story.) Reports are released to the press, or leaks reach reporters of not-yet-released reports. Someone gives a speech, a press conference is held or one of the thousands of press releases arriving on an editor's desk gets attention. Or someone makes a phone call. It is possible that a concerned citizen's letter to an editor plants the seeds for a story, but the usual initiation of coverage is an involved source. Unfortunately, there seems to be a paranoia, especially among government sources, about looking pro-pesticide. The attitude seems to be that any coverage leads to unnecessary concern by the public. Scientists are afraid to release information to the press and therefore don't, leaving the flow of information to accidents and leaks.

I do not want to misrepresent this as a formal survey because in preparing this paper I have only reviewed the existing literature and talked to an unscientific sampling of reporters who cover pesticide stories. My judgement is that their comments are fairly representative. Journalists I spoke with generally cited the same sources for pesticide stories. They look for someone from a responsible agency who can talk, a company representative, a victim or a witness. Primarily they depend on authorities. They go to local experts for translations of jargon, background information and better understanding, and they are using hot lines such as SIPI's Media Resource Service as much as possible. They report that most pesticide related stories are breaking stories--news happening right now-- not in depth follow ups or features. They also list the usual constraints of time and available sources. Reporters say they feel their coverage has made the public aware of pesticides and maybe a little more cautious--"scared of the stuff." They complain of poor relations between scientists and journalists and lament that they "don't get any respect." They're not sure how to improve the science/ media relationship, but they want the experts to be willing to talk more. Everyone I spoke with rates his or her publication's or station's coverage of pesticides as accurate and fair. They think they're doing the best possible job. Sadly, a few report they are beginning to share the public's lack of faith in the industry's openness to change. Along with the public, reporters are asking, "If it's safe why are you telling me about it?" or, "If it's dangerous why is it in my universe?" Standing in the middle sounds unresolved, too technical, too hard to sell to either editors, or the public.

Very little research exists on how the public use the mass media for risk information. The psychology literature examining what affects nonscientists' perceptions of risk is useful, but mass communication and media scholars are just beginning to gather data in this area. Sharon Dunwoody at the University of Wisconsin is finishing a content analysis of media risk stories to see how such information affects individuals' risk perceptions.

In New Jersey Peter Sandman and a group of researchers from Rutgers University are involved in a far reaching project examining that state's environmental risk reporting. Their work so far has

found, most importantly, a lack of reporting about environmental
risk (1). Experts involved in the New Jersey study suggest reporters
are more at ease covering environmental politics than environmental
risk. Other important findings from this research effort include:
an unmeasured "feeling" that when risk is reported in New Jersey
newspapers, it is more alarming than reassuring; reporters rely primarily on the government (over half of the cited sources were local,
state or federal government) and industry sources who were found to
be the least likely to pay attention to risk and were the leading
sources for risk-denying; risk is assessed by journalists in terms
of extremes rather than quoting intermediate or tentative positions;
and, bias, when it occurs, results from overreliance on a single
source or a failed effort to translate jargon into lay terms. One
of the key recommendations of this report for editors and reporters
is more reliance on expert sources who are uninvolved in a particular
story. It is reassuring that this New Jersey study did find experts,
who were among the most likely to assert risk, being cited in over
one fourth of the articles analyzed.

Prominent in laying the foundation for how people deal with risk
is Paul Slovic. The conclusions and recommendations Slovic, with
Baruch Fischhoff, Sara Lichtenstein and other risk perception experts
offer are very important to those who are now focusing on how media
might inform and impact public perceptions. (See for example
Acceptable Risk (2).) What does it mean to consider media's role
when the subject demands attention to values, beliefs, and issues
rather than events? Issues of fact and issues of value are not
generally debated in newsrooms; the two are clearly separate to the
journalist. However, as we all know, a clear separation is not
always possible. Judgement calls are made regularly by experts
when confronted with perceived vs. objective risks. The public or
nonexperts bring an even more diverse set of viewpoints to the analysis. Value judgments, uncertainties, complexities, fallibility,
and even inconsistencies all play a part in assessing acceptable
risk. Tell that to a fact-finding journalist and you're in trouble.

MEDIA'S ROLE

What impact do media have? There are five effects generally cited.
The first impact is persuasion, but there is much misunderstanding
about media's powers to persuade. What is hard for most non-media
professionals--and some media pros as well--to accept is that regardless of the issue, mass media cannot convert their readers,
listeners or viewers. Mass media reinforce the existing attitudes
of the audience; they are not a force for change. Media cannot be
used to force conversion. What can be done is to attempt to understand which existing attitudes should be maintained in working toward a goal. What beliefs and opinions forming the status quo
are only slightly different from the desired new view? What current
audience attitudes can be moved in the direction that would be acceptable to the established goals? Psychologists call this ability
to maneuver around an attitude and create a new view a "latitude of
acceptance." In mass communication theory, it is also acknowledged
that mass media can be used to guide an audience toward a new attitude that is rooted essentially in what they (the audience or

message receivers) have already accepted, what they already believe. For example, if people are already fearful of pesticides but only in reference to plant accidents, fires or other disasters, consider the potential usefulness of such fears to promote the safe use of chemicals. Don't deny the fear. Use the existing caution induced by fear to establish the desired new attitude or behavior.

Media's second impact comes in the form of agenda-setting, which means that editors and reporters respond to what they think the public is interested in, not what they think the public ought to be interested in. It is probably true that media are often ahead of their audiences and are in reality "bringing the public along," but it is nonetheless important to understand what is selected for coverage and where it shows up. If pesticide risk is to be an agenda item, editors and reporters must be convinced that the public considers pesticide risk an important issue. If neither the public nor editors list such risk among their concerns, the story will never see the light of day...until there's an attention getting disaster. The industry it seems is faced with a serious educational task. Recall that nuclear power was not much of an agenda item until Three Mile Island (TMI). Pesticides--not even the manufacturing sites--were not on many lists until Bhopal. As do journalists, the industry waited for a disaster before attending to the issue. It is ironic that both media and industry failed to respond to what was clearly a concern for the public (3).

The risk stories to which media seem to have attended of late are the dangers of smoking, the risk of getting Acquired Immune Deficiency Syndrome (AIDS), and the problems of hazardous waste disposal. Those news items are for the most part the result of legislation, industry press releases, advocacy group campaigns and real events covered by reporters. Perhaps now that Caesar Chavez, President of United Farm Workers of America, has undertaken a strategy which calls more public attention to pesticide use in America, editors will begin to add pesticides--perhaps even pesticide risk--to their agendas. That of course casts industry in the role of respondent rather than initial source. Some sources have learned the lessons of media power better than others.

The third area of media impact is in presenting norms; media attempt to reflect our arts and social customs. There is no question that media homogenize culture in a regretable way. There are those who devote their lives to improving this content and form dilemma. The fourth and fifth areas of media impact are also less than desirable. Media are held responsible for modeling--fashions are straight out of "Miami Vice"; our children think in Smurf. And, media induce apathy. Until more can be done with the interactive potential of TV, media are a passive experience. And worse, that passivity is satisfying the total experiential need. Even the most ardent advocate of television reacts with horror to the statistic showing the percentage of people who rely solely on the six o'clock news for all of their information. A 1982 study indicated that 41 percent of the American public rely on television as their only source of news and that 53 percent feel television is more credible than other media (4). It would be of immense value to society if simply viewing the tragedy of farmworkers' exposure to pesticides initiated actions that led to real safety measures. It would be useful if reading a story about

the dangers of pesticide residue in produce resulted in an overwhelming consumer demand for better labeling and rethinking of costs
vs. benefit. But in fact, such programs and stories are experienced
as a whole--the readers/viewers/listeners feel they have the beginning, the middle and the end of the story, and there is no need to
act further. The audience is either lulled or numbed into passivity
by mass media.
       These general media impacts have been understood for some time
(5). What we can say about specific perceptions of the audiences
reached by media is less clear. But before we move whole-heartedly
into researching what the results of the message on the receiver
might be, it seems to me more effort can be put into improving how
sources are delivering the messages, and what is being said. The
relationship between sources of science information and the translators who send the messages is not yet the best it can be.

MORE PROBLEMS

The usual charges when criticizing media coverage of risk issues do
not differ so greatly from problems cited for media coverage of
science and technical issues in general. First, the complexity of
the information requires science literacy. Accuracy is difficult to
achieve. Science reporting in general is incomplete. Assessing
risk requires good judgement and, as St. Petersburg Times president
and editor Andy Barnes points out, some of today's new journalists
don't even know the norms (6). The second category of charges suggests reports are biased, superficial, sensational, negative, distorted, and generally lack objectivity. The third general charge is
that the "good" news story is always bad news.
       In discussing media constraints when covering or presenting
science, June Goodfield (7) also points to financial pressures,
especially in television. Goodfield sees the sword of Damocles
hanging over producers, threatening to cut off funds. Any documentary or expanded programming requires financing and sponsorship.
Intervention in terms of content is not so much the issue here as
is initial funding and eventual marketability of the idea.
       I don't find journalists denying that accurate science report-
ing is difficult. Those in media are working to provide better training for future science writers, and better support resources such as
SIPI are becoming more available. I do not take seriously charges
that editors and news directors have a "startle-amaze-amuse-them"
mindset, and insist on simplicity and negativism. Such criticism
reflects an unwillingness to understand or respect the process and
constraints of mass media. Neither do I accept the charge that somehow the public's mindset is to blame for missing or mis-understand-
ing. The public knows scientists can be wrong. TMI for example, or
a failed shuttle launch, do much to reinforce that attitude. And,
experts have noted there is often much to be learned from the general
public's common-sense wisdom. Rather than accuracy, objectivity, or
motivation, what seems the more serious problem for journalists and
science sources seems to lie in basic values.
       Several very different "world views" between journalists and
sources have been noted. The comparison (see Table 1) foretells
difficulties. For example, whereas journalists search for facts and

are triggered into action by events, scientists, as do most sources,
look for truth and confront issues. Whereas journalists report what
happened (a yesterday focus), scientists see time as forever.
Sources are allowed to feel strongly about their work, whereas journalists are bound by a creed of objective, unfeeling rationality.
Reporters have a commitment to the process of news gathering regardless of results, whereas sources are clearly focused on the end
message--what appears in print. Journalists record events and balanced quotes in a point by point, one fact on another (atomistic)
style, whereas a source has a more coherent vision. And so on.

Table 1	
Journalists	Sources
facts/events	truth/values/issues
yesterday	forever
rationality	feeling
peers	audience
process	result
fast	timeless
atomistic	coherent
interesting	important
Source: Sandman, April 1984, lecture at University of Tampa.	

Each of these sets of views offers very different integrities, and
very different sets of values.
     Television of course futher complicates the attention to dif-
ferent values because of its reliance on good pictures. TV is primarily a visual medium. As Roger Peterson of ABC news points out,
"Bophal was dynamite television (8)." The impact on innocent people
provided what television needed--shocking pictures full of emotion.
In comparison, TMI was not such a good picture story, but the immediacy of impending diaster made up for the lack of visuals. The
problem in covering pesticide risks is similar to TMI in that there
is no way to really show ethylene dibromide (EDB). Television can
show people reacting to or talking about EDB dangers, but it's
simply not a good picture story. Unless, of course, there's a fire.
     To many of us the more serious difficulty is that it seems im-
possible to get on the air with the issues between crises. The long
term problems don't get coverage. Series or in-depth features are
few and far between, and then, some journalists bemoan the task of
finding angles that make interesting pictures. Regardless of content, television can still be counted on to choose good pictures
first, and immediacy is primary.
     It may be frustrating to the experts, but reporters simply do
not have the time to understand it all. News on deadline is not a
think piece. It's enough to "develop a feel for what's dangerous,"
and "find an expert and pick his brain for a translation." And that
means the front page story is more likely to be incomplete while the
Sunday supplement story offers far more background and information.

Any attempt to extend information requires looking to documentary or
news analysis treatment.

THE MEDIA PROCESS

Some of the best science writers get their leads from reading scientific journals and publications. They read and review reports,
and check with experts in an effort to understand the process and
findings of research. But, pursuing such stories--those not pegged
on a specific event--is a luxury in the world of media. There is a
difference between a publication of record and a thorough examination
of an issue. And, there's a difference between the straight science
story--that's the usual bread and butter story--and an issue story
or a science policy story. Reporting pesticide risk is an issue and
policy story, unless there is a fire, a spill, or another clear
newspeg.

Although the earlier described newspeg is core to the Who, What,
Where, When, Why and How, the standard formula for newsworthiness to
a journalist also includes proximity, prominence, unusualness, human
drama, consequences and immediacy. An editor and the reporter make
these judgement calls, and then, they turn to sources. "If I can,
I will avoid the PR person...I want to go to the scientist who did
the study, the person who made the decision," any good reporter will
tell you. A good public relations person knows that original sources
are imperative and will act as a liaison for media to that quotable
expert source. Sandman's research in New Jersey tells us a lot
about who journalists use as sources (9). It is useful to note what
these sources say about risk coverage.

Environmentalists label the results of risk coverage in terms
of "who cares?" and "so what?" In general they feel the coverage is
so poor and unarousing that only an occasional headline has any
impact. Industry PR representatives seem dismayed with the editing.
The stories are so full of holes that the public can not possibly
make intelligent decisions. Experts feel reporters do not have even
the basic understanding to explain the information, but that
the inaccuracies reflect a lack of knowledge rather than bias.
Reporters themselves admit to not asking enough questions but are
overwhelmed by the lack of standard assessments of risk and uncooperative sources.

Journalists want non-adversary interpretors, someone to help
formulate the question as well as communicate with ease for public
consumption. They're asking for "user friendly" expert sources, not
fact sheets, but real people who can be interviewed and quoted. I
hear repeatedly from science and environment reporters about the
difficulty in finding "leading experts with no ax to grind." In
Scientists and Journalists Dunwoody cites research pointing to three
major credibility factors a journalist looks for in selecting a
source for a pesticide story. Journalists look for a source with
mainstream status, administrative credentials, and previous media
contact (10). These sought after traits seem far more manageable, I
would think, than Goodell's description of the fittest visible
scientist for media as one who is relevant, controversial, articulate,
colorful and reputable (11). SIPI has a list of 150 experts who
might respond to a reporter's inquiry in a pesticide risk story.

Of these, 70% are affiliated with universities or research centers,
10% are with environmental groups, 10% are employed in government
agencies and the rest are in a variety of other positions (12).
    It is ironic that reporters do assume bias on the part of every
expert. That's probably a healthy suspicion that can only lead each
assignment to include more sources. Expert bias and the well
founded belief that experts can't talk to people leave journalists
more than content with their roles as translators, sometimes floun-
dering around, but doing their job as they see it. Multiple sour-
cing--more translators--may even offer better long term understanding
for journalists. The more complicated obstacle to overcome seems to
be a reporter's difficulty in accepting that an expert may truly not
know the needed answer. Too many journalists have routinely assumed
a cover-up or general unwillingness to provide information when the
answer is, "I don't know." Not knowing may not be so difficult for
a long-range-thinking scientist to admit, but "I don't know" is near
impossible to get past an editor.

THE EDB EXAMPLE

For the Tampa Tribune, EDB began as a local story. Two counties in
the newspaper's circulation area were among the first to be cited for
contamination with the pesticide. Tribune reporters, particularly a
very enterprising young woman from the newspaper's Lakeland bureau,
interviewed state health officials, government agencies, local mil-
lers and farmers, Florida Citrus Mutual, and scientists before they
even began to note the importance of the breaking story. "It was
tempting to wonder if this pesticide, which was widely used for more
than two decades on groves, fruit and grains, wasn't simply the
latest "'chemical of the month'," wrote Tribune state editor Bill
Gueskin (13). The Tribune ultimately saw EDB as "both a frightening
danger to public health and a vivid example of the difficulties
government agencies encounter in protecting citizens." That was in
early 1984. Almost exactly two years later the same newspaper's
Lakeland bureau business writer has no difficulty reporting, "The
fumigant [EDB], used to destroy fruit-fly larvae, is essential in
maintaining Japanese markets for Florida grapefruit (14)." Again
the sources for the current stories were Florida Citrus Mutual, the
EPA and now, Great Lakes Chemical Corporation.
    The EPA commissioned study on how the public received the mes-
sage that EDB was unsafe (15) used content analysis of 50 newspapers, news programs, national press stories and weekly magazines to
arrive at a conclusion that indicts "macro-risk and micro-risk"
perceptions on the part of EPA's specialists vs. the public's vision.
According to this study, neither the EPA nor the public were in error
about risk perceptions. The difference between perspectives (macro
vs. micro) caused a barrier. Nonetheless the report characterizes
the public during this six month period in 1983-84 as "confused and
antagonistic (16)."
    What NBC's chief censor Ralph Daniels knows, this EPA document
does not tackle: there is no mass audience in spite of the mass
media. Individuals view programs and that presents endless problems
for people like Daniels whose job it is to be not only accurate, but
a respectful guest in someone's living room. Reporters translating

government agency messages about EDB were only trying to provide
understandable information for each media consumer. No editor or
reporter set out to create a science fiction monster. If the New
York Times or Wall Street Journal have superior coverage of a story,
it reflects lessened constraints and more expertise on the part of
their journalists, not a macro view as opposed to a local newspaper's micro view.

EPA's EDB report clearly notes that the pesticide had been ex-
empted from all regulations as long ago as the mid-fifties and had
been "in our universe" since the forties. Why then should it seem
amazing that the public some thirty to forty years later was un-
soothed by assurances that the risk from this chemical was only long
range and chronic? It did not take television's image of grain ele-
vator workers overcome by toxic effects, being rushed to a hospital
(where it is later reported they have died) to create alarm. The
EPA's own, original, untranslated communications were adequately
fear arousing.

The irony is that regardless of media "sensationalism" the
public, at least in Florida, apparently remained disbelieving, un-
interested and generally uncooperative with local task forces. Some
light might be shed on this public reaction if sources for the EDB
study are considered. The Miami Herald for example, as did other
state newspapers, relied on the State's Department of Agriculture,
industry sources such as Monsanto Chemical Company, and interviews
with citrus farm owners for sourcing. With rare exception, reporters
ignored workers and consumers. Even in the wake of a recall order,
three months into the EDB scare, all was relatively quiet in the
state. And nationally, media all but dropped the issue, not when
public anxiety declined or when EDB disappeared, but when the press
releases and press conferences stopped. The "event" had been covered
and it was over. There were no rioting or hysterical hoards. Even
the EPA report notes without insight that NEWSBANK stopped indexing
EDB articles after February 1984. As with nuclear anxieties, economic scares and other over-sized fears, the public accepted media's
closure on the issue, or became numbed (17).

THE PROBLEM SUMMARY

The more general outcome of this information complexity, lack of
expertise, no clear and immediate answer, and multiple, media-
process constraints is the reporters' ethic that admonishes them to
err on the side of public safety. And, as Paula Lyons of WCVB-TV in
Boston sums up, if media are making people more afraid--more
careful--"that's okay (18)." Accusations of being alarmists concern
reporters far less than being caught missing the story. And to cover
the risk story, a bit of fear or a measure of alarm might be the only
news peg.

SOLUTIONS

One key to improving communication between journalists and scientists
is a clearer understanding of the media process. As stated earlier,
the story idea is generated, usually by a press release, an event, a
study or report, but a leak can substitute as the trigger. The idea
is evaluated and assigned by an editor. The story is covered by a

reporter or team of reporters. (And here is where scientists as
sources play a critical role.) Space and time are allotted. The
story is edited and processed.

The overshadowing professional demands at each step of the process are honesty, accuracy and fairness. The mass communication
demands are clarity, readability and style. These demands for delivering information to the public need not conflict, but in covering science and the complexities of risk, they often do. The dilemma
is real but can hopefully be resolved.

Dr. Vincent Covello, Director of the National Science Foundation's Risk Analysis Program, has developed a list of ten questions
he believes a reporter should ask to assess risk (19):

1. What is the probability that people might be harmed and to
   what degree?
2. How much of the assessment of risk is based on assumption
   or guesswork?
3. If there is an uncertainty in the data, do the conclusions
   reflect that?
4. Does the study consider the number of people exposed to
   the problem?
5. What are the study's limitations?
6. Do the researchers consider such things as individual
   sensitivities, exposures to multiple hazards and cumulative
   effects?
7. Are all the scientific data open to the public scrutiny?
8. Does the analysis distinguish between voluntary and involuntary exposure?
9. Is the process of doing research kept separate from the
   process of making policy decisions?
10. Who provided the funds for the study?

Rather than Covello's approach, however, the norm now is to find
risk--if included at all--in a single paragraph answering:

1. How much is there?
2. What's the standard?
3. What's the health relationship?
4. Who objects or disagrees?

Either formula for good coverage requires digging, phone calls and
understanding. None of this is initiated using either formula if
risk or science issues are not assigned as stories. And nothing gets
covered without access to good sources.

Burson-Marsteller and Hill & Knowlton, leaders in the public
relations (PR) industry, are impacting the public's perception of
pesticide risk, and they are premier in knowing how to use media.
Perhaps the American Chemical Society should use this public relations expertise just as Union Carbide and Dow have, but I do not
think such reliance on these giants of PR is the only or even the
best solution. Uncovered PR-generated articles are neither journalists' nor the public's favorites; such stories do little for long
term creditility.

Better suggestions come in a very recent report on risk communications, again by Covello, listing communication problems with
the message, the channel, or the receiver (20). In other words,
something goes wrong with understanding and assessing what has been
said or what has happened, something goes wrong because of the risk

assessment experts, something goes wrong because of the media process, or something goes wrong because of characteristics inherent in the mass audience--the public.

Let's consider how well media might fare if the tasks of risk communication are as outlined in Covello's report:

1. The task is treated as one of information and education.
   Some mass media do well as sources of information, but there are format and process constraints on how messages will be presented. There are serious concerns about media's educational role.

2. The task is seen as behavior change and protective action.
   Media rarely change behavior, particularly for strongly held beliefs. It is necessary to look to the advertiser's model for persuasion and understand mass communication theory, especially in regard to fear appeals for success in this type of communication.

3. The task is to send disaster warnings and emergency information.
   Experience and conditioning indicate media have strengths in this role, but sources must have a good track record on these occasions. Media are relinquishing control under prearranged plans for such communications.

4. The task is one of joint problem solving and conflict resolution.
   Media may play a supportive role in, for example, announcing public hearings, but in an effort to balance a story, confusion or undue alarm is often generated.

Rather than hiring Hill & Knowlton or Burson-Marsteller, Covello's report suggests a very useful list of what communication can do to effectively inform people about risk. Effective risk communication requires simplicity; relevant, personalized comparisons; an understanding of the audience; complete honesty; and a perspective that acknowledges political and ideological conflicts. It is also useful to remember all media channels, such as pamphlets and alternate video uses, instead of an exclusive focus on commercial television and newspapers. (Beyond news coverage of pesticide risk, someone needs also to look at how entertainment programming represents such issues.) If sources followed the offered guidelines, media's translation and transmittal would be less suspect as the culprit for public misconception.

A ranking of the most popular topics presented at the 1977 AAAS annual meeting listed "world food losses to insect pests" as in the top ten for experienced science journalists, who ranked the story at #6. Scientists did not rank the topic as popular (21). In 1984 the annual Associated Press poll placed the Bhopal disaster at #2 in the top ten stories of the year. Many reporters still use a time reference labeled TMI to Bhopal. In 1984 the only stories ranked as tops by national editors and broadcasters that came close to being science related were the Ethiopian famine (#6) and heart transplants (#8). The Mexico City gas explosion and space shuttle satellite retrieval were both ranked among the second top ten. Five out of twenty top stories for one year in somewhat of a science category is actually a dramatic rise in attention to science news.

In 1985 the top ten stories included an earthquake and a volcano in addition to a famine and the AIDS epidemic, but economic and

political stories were again dominant. The obvious note to make is
that diasters, natural and unnatural, are considered the best news.
A 1982 Gannett poll showed that newspaper readers rated natural disasters and tragedies as most popular, with local and national economic stories as second and third. Stories about the environment,
energy or conservation ranked seventh. There are no risk stories in
the top ten, not the risk of economic depression or even the risk of
an air diaster. The news is the actual stock market crash, or hi-
jacking. Even a meeting is more likely to be considered news than
is a risk story.

It is no wonder most scientists prefer magazines to other news
media. Not only are magazines somewhat similar to the more familiar
journals, but the magazine medium is less hampered by the constraints
of newspapers or electronic news media (22). And now in the 1980's
there is an explosion of popular science magazines fueling the pub-
lics demand for more information. That may be a factor in the fact
that newspapers have increased the number of pages devoted to science
coverage (23). Magazine articles can be longer features, more re-
flective and verified. Whether working with a magazine writer or a
reporter from another medium, Barbara Gastel's Presenting Science
to the Public (24) offers a good primer. More seasoned veterans of
encounters with media professionals may find it useful to simply make
the time to talk one on one--get to know these people as people. Or,
participate in opportunities to encounter journalists at events such
as Boston University's one day seminar on media coverage of public
health issues, or Northwestern University's seminar on science and
health reporting, cleverly titled, "Risky Business." Far more attention has been paid lately by universities, the AAAS, and professional
organizations such as ACS to getting science specialists and media
professionals together. The common goal is to have a more informed
public able to make more informed decisions.

It seems to me the public feels trapped by a system that tells
them pesticides are essential to adequate food production and at the
same time suggests pesticides are carcinogens. They would like to
believe the scientists who assure them that Temik is not harmful,
but they fear the worst. Although disastrous physiological consequences have not yet resulted, there is an uneasiness in the land
and more frequent reference to Rachel Carson's Silent Spring (25).
Mothers no longer tell their children to wash the fruit and eat that
vitamin-rich peel; now caution and uncertainty have promoted avoid-
ing the skin where "all those pesticides concentrate." Public perceptions appear to range from paranoid to cavalier, although little
real measurement data exists.

For the public, mass media are the dominant sources of information about risks. Media set the agenda, shape and frame the reality
by accurate and timely treatment through coverage or non-coverage,
and reinforce what the public already believes. Policy and public
opinion leaders respond to media coverage and in turn try themselves
to influence media content. The media are critical in any risk communication effort.

The majority of expert media watchers do seem to feel media have
done a good job in reporting risk, although excesses, errors, and
bias are noted. There is general agreement that to supress media
coverage of risk would deny the public's right to know and violate

the freedom and responsibility of the press. I'd like to summarize
what I've tried to say by offering a list of what I believe can be
done to make better media coverage and indirectly, clearer public
perception of the pesticide risk issue, more possible:

1.   Understand how public perceptions are formed. Do you have
     public relations campaigns to inform the public? Media
     are not totally, if at all, responsible for public per-
     ceptions.
2.   Don't blame the press for actions taken by others as a
     result of information provided in the media.
3.   Understand that the reporter covering the respective risk
     story may not necessarily be a science journalist; the
     general assignment reporter will require much more patience
     and translation and even science writers may need assis-
     tance in developing better thinking skills.
4.   Drop unessential jargon and clearly define science terms.
5.   Provide as much information as available.
6.   Learn to stop hating the press and respect their profes-
     sionalism. Forget and forgive their past mistakes.
7.   Cooperate. It can only help. Remember that reporters are
     also apt to have an unforgiving memory for deliberate or
     perceived omissions and lies.
8.   Don't confuse alarm with sensationalism.
9.   Don't expect media to do the slow buildup stories. And
     don't expect the seemingly necessary follow-up story. (Do
     encourage such coverage and help make it happen.)
10. Don't expect television to provide what a Wall Street
     Journal article might provide. Know which medium to lean
     on for what.
11. Allow reporters their humanity. Journalists care very
     much about consequences and are generally very active mem-
     bers of society. Some of them will even confess to resent-
     ing their editors' demands for inhuman impartiality, dis-
     interest in issues and commitment to being uninvolved in
     the news. A job that requires a separation of facts and
     values is neither an easy one, nor one with which scien-
     tists should have difficulty empathizing.
12. Try encouraging compassion rather than cynicism. In the
     November 1983 issue of The Quill Gene Goodwin noted the
     substantial newsroom lore creating that hard-nosed, macho,
     dispassionate person demanding that two quotable facts be
     handed him or her, and now. Without superiority or sub-
     terfuge, try treating reporters as people who care as much
     about accurately informing the public about risks as you do.
13. Understand the process and constraints of mass media.
14. Offer statistical comparisons for a reporter attempting to
     translate and put into perspective a risk message. Cer-
     tainly a statistical comparison is more relevant and more
     sensible than leaving someone to ask, "Do you want worms
     or EDB in your pancake mix?"
15. Do not wait to release weighty background papers, defini-
     tions and fact sheets in the middle of a breaking story.
     That is at best too much too late. Those resources and
     that understanding should be available and reinforced on a

continuing basis. As the story breaks, references then
become a refresher rather than an impossible task.
16. Remember that being on deadline does not allow time for a
lengthy explanation or "the history of..." Try to find out
specifically what the reporter needs to know, then provide
answers (or help restructure their questions) without
lecturing. Their impatience is not with the subject;
they're facing white space or dead air and the clock's
ticking.
17. Encourage and participate in workshops or other efforts
to bring scientists and journalists together for mutual
"training" sessions.
Overall, rather than post disaster planning, I recommend a
better understanding of the public's fear and confusion, and a more
cooperative posture toward media. If the public is uninformed, media
can play an important role in providing information. If the public
is misinformed and overly fearful, media can be useful in turning
that fear into positive action.

Literature Cited

1. Sandman, P.M.; Sachsman, D.B.; Greenberg, J.; Jurkat, M.;
Gotsch, A. R.; Gochfeld, M. "Environmental Risk Reporting in
New Jersey Newspapers," Environmental Risk Reporting Project,
Industry/University Cooperative Center for Research in Hazardous
and Toxic Substances, New Jersey Institute of Technology,
January, 1986.
2. Fischhoff, B.; Lichtenstein, S.; Slovic, P.; Derby, S.L.;
Keeney, R. L. "Accpetable Risk";Cambridge University Press:
Cambridge, 1981, pp. 28-35.
3. Allman, W. "Staying Alive in the 20th Century"; Science 85,
1985, 6, 31-41.
4. Hiebert, R.E.; Reuss, C. "Impact of Msss Media"; Longman:
NY, 1985.
5. Sandman, P.M.; Rubin, D.M.; Sachsman, D.B. "Media"; 3rd Ed.
Prentice-Hall: Englewood Cliffs, NJ, 1982.
6. Barnes, A., personal communication.
7. Goodfield, J. "Reflection on Science and Media", American
Association for the Advancement of Science: Washington, D.C.,
1981, AAAS Pub. No. 81-5.
8. Peterson, R.; Invitational Symposium on Environmental Risk
Reporting, 1985.
9. Sandman, P.M.; Sachsman, D.B.; Greenberg, M.; Jurkat, M.;
Gotsch, A.R.; Gochfeld, M. "Environmental Risk Reporting in
New Jersey Newspapers", Environmental Risk Reporting Project,
Industry/University Cooperative Center for Research in Hazardous
and Toxic Substances, New Jersey Institute of TEchnology,
January, 1986.
10. Dunwoody, S. In "Scientists and Journalists: Reporting Science
as News"; Friedman, S.M.; Dunwoody, S.; Rogers, C.L., Eds.;
Free Press: New York, 1986; p. 7.
11. Goodell, R. "The Visible Scientist"; Little-Brown: Boston,
1975; p. 18.
12. SIPI, personal communication.

13. Grueskin, B., "Stories Uncover the Problems with EDB". Tampa Tribune, February 4, 1984, D, p. 1, Col. 1.
14. Garmond, G., "Feds OK Seasonal Use of EDB". Tampa Tribune, February 14, 1986, E, p. 2, Col. 5-6.
15. Sharlin, H.I., "EDB: A Case Study in the Communication of Health Risk," Office of Policy Analysis, U.S. Environmental Protection Agency, January, 1985.
16. Sharlin, H.I., "EDB: A Case Study in the Communication of Health Risk," Office of Policy Analysis, U.S. Environmental Protection Agency, January, 1985, p. 13.
17. Sandman, P.M.; Valenti, J.M. Bulletin of the Atomic Scientists, January, 1986, 42, 12-16.
18. Lyons, P. Hazards or Hype. Video Tape Seminar, "Public Health and the Media". Boston University: Ketchum Public Relations, 1984.
19. Covello, V. Science and Health Reporting: Risky Business. Video Tape Seminar. Northwestern University, Medill School of Journalism, 1985.
20. Covello, V.; von Winterfeldt, D.; Slovic, P. "Risk Communication: An Assessment of Literature on Communicating Information About Health, Safety, and Environmental Risks". Draft Preliminary Report to the EPA, Institute of Safety and Systems Management, University of Southern California, January 11, 1986.
21. Dunwoody, Sharon. In "Reporting Science: The Case of Aggression". J. Goldstein, Ed. LEA Publisher: London, 1986, pp. 75-76.
22. Dunwoody, S.; Scott, B. Journalism Quarterly, 1982, 59, 52-59.
23. SIPISCOPE, Scientists' Institute for Public Information, 1984, 23, 1-16.
24. Gastel, B. "Presenting Science to the Public"; ISI Press: Phila., 1983.
25. Carson, R. "Silent Spring"; Houghton-Mifflin: NY, 1962.

RECEIVED August 21, 1986

PANEL DISCUSSION

Chapter 14

Summary and Discussion

Robert E. Menzer

Department of Entomology, University of Maryland, College Park, MD 20742

In the overview to this book Dr. Alvin L. Young presented a striking
item of information: when one considers the investment in pest control versus the gain in productivity that results in the United
States, a $9 billion estimate of the value of pest control is ob-
tained. In spite of that we are still losing 30% of our total agricultural production to pests in this country, and an even higher
percentage on a world-wide basis (1). This fact embodies the challenge in minimizing the risks of using pesticides: if we could
somehow increase by only a few percent the amount of pesticides
that reach the target and expand our knowledge base to support effective use of pesticides, we could save, in the first place, bill-
ions of dollars, but we would also be minimizing to a significant
extent the risk of pesticides to the environment.

Minimizing Risk by Understanding Toxicology

The U.S. Environmental Protection Agency (EPA) requirements impose
on us a certain understanding of toxicology before a pesticide can
be registered. Dr. Ray Cardona has very effectively presented the
specifics of the guidelines for developing toxicology information
for registration which were adopted by EPA several years ago. The
question is raised, however, whether the rigidity, or lack of flexibility, of those guidelines prevents or in any way inhibits the
application of new technologies, including biotechnology, to the
development of the new compounds which will be needed in the future
to lower or minimize the risk of pesticide use.
    The question of our focus on acute versus chronic toxicology
is a problem area in the continued use of pesticides. We have put
our major research and data-gathering emphasis in the past on acutely toxic chemicals and acute toxicity problems. When we started
considering chronic problems, as with the chlorinated hydrocarbons,
we discovered major problem areas. That led to quite significant
changes in our perceptions of the ways we ought to be using pesticides in general, certainly insecticides specifically.
    A question which is perhaps just as significant as the switch
from concern about acute toxicology to chronic toxicology, is the

0097-6156/87/0336-0168$06.00/0
© 1987 American Chemical Society

coming emphasis on the question of interactions. Very little is
known about interactions of pesticides in the environment, with
each other or with endogenous chemicals.   Dr. Raymond Yang gave
several examples of some interactions that could be harbingers of
what may be occurring in the environment that have not been recognized.   The need for research to address this matter is great.

Several chapters present information on the possibilities for
simulation and modeling, both in terms of understanding the data
which have already been accumulated, but also in terms of developing
new techniques for discovering new compounds, based on information
which has been obtained from structure-activity relationship studies.   If we can predict the effects of compounds before we actually
put them into the field, obviously the risk of the toxicology of
these compounds to both agricultural workers and consumers of the
products, and to the environment, will be minimized.

Minimizing Risk by Understanding the Pest

Understanding the biochemistry and physiology of the pest and exploiting that as the search for new compounds proceeds is receiving
much attention today, as well it should.   In the past, that generally has not been the case.   It was just 20 years ago that we talked
about the "spray and count" pest scientists, who really knew very
little about the habits of the pest;   all they were really concerned
about was how many would die given a specific dose of pesticide as
applied in the field. We have now progressed beyond that approach,
as several chapters point out.

The interaction of the pest with its environment must be taken
into account in order to use pesticides optimally. We have, of
course, considered the environment in development and use of pesticides, but frequently we have been more concerned with the non-target species which also happen to inhabit the pest's environment and
not so much with target species-environmental relationships.   If we
knew more about the way pests behaved in the environment, we would
be able to direct pesticides to their targets better and in a timely
manner that would give more effective results.

In addition, we are increasingly concerned about pest resistance to pesticides, which has been a matter of some concern for a
long time.   The intriguing idea that we might be able to exploit
and overcome the pests' resistance mechanisms was discussed.   By
taking some of the pests' traits which are negative for resistance
into account, one might "trick" the pest as it develops resistance
to one pesticide by using another compound in conjuction with it
that would exploit these negative resistance factors and thus overcome the resistance. A recent National Academy of Science report
(2) has emphasized the need to focus more resources on understanding
and overcoming the resistance problem.

Minimizing Risk by Understanding the Chemical

In order to minimize the risk of pesticides by understanding the
chemicals, Dr. James Seiber maintained that we must understand the
chemodynamics. We have only just begun to consider factors such as
water solubility, octanol-water partition coefficients, soil

adsorption constants, etc., to understand the behavior of existing
compounds in the environment and to try to design better compounds
in the future. If we had really paid attention to the chemical and
physical properties of pesticides and stopped to think about some of
the things that should have been obvious to us, and were obvious
once we did think about them after the crises had erupted, we would
probably not have used certain compounds under some circumstances.
We had a very dramatic example of that in the case of the use of
aldicarb on Long Island. Following extensive use of this highly
toxic insecticide for the control of the Colorado potato beetle, it
was noted that the material readily leached through the coarse sandy
soils characteristic of Long Island and finally reached the ground
water. Contamination of the only source of drinking water for the
area must be classed as a severe environmental crisis.

Another illustration of our failure to use existing knowledge
is presented by Dr. Wyman Dorough in the discussion of mammalian
and other animal metabolism of pesticides. Much of what we know
about metabolism of pesticides in mammals and other organisms was
not predictable prior to the development of the chemicals. We have
now come to the point, however, where we can make some predictions.
Although our regulatory approach requires us to actually do the research, we must also think about the implications of what we are
learning. Then we could apply our knowledge to the design of new
compounds and for minimizing the risk of both old and new compounds.
Information gained about comparative toxicology and comparative
metabolism of these chemicals will minimize the risk to the ultimate
non-target species, man.

Finally, the ultimate benefit of the accumulation of this
chemical information is to apply it, using some of the really ele-
gant new computer techniques now available, to the design of new
compounds that will nicely fit the specific biochemical target sites
of pest species, while at the same time not fitting targets, or
similar targets, of non-pest species. Thus we can achieve what
might be called the ultimate selectivity.

Minimizing Risk by Understanding the Hazards of Pesticides

One might consider two separate populations of people who need to
understand pesticide hazard: those who are occupationally exposed
and the general public. The hazards of pesticides, and specific
steps taken to minimize them, have been the focus of environmental
activist groups, farm workers' unions, and the Cooperative Extension
Service. Great strides have been made toward minimizing exposure
of workers to pesticides, both by providing protective clothing and
improving the design of application equipment, and by information
programs designed to encourage workers to improve their own work
practices to minimize their own exposure.

Communication of the realistic hazards of pesticides to the
general public has been more difficult. Dr. Ronald Hart's statement that risk assessment techniques must be upgraded to use them
more effectively, is an illustration. The approach using several
orders of magnitude difference in numerical risk assessments simply
is untenable when one tries to apply this in practice and communicate the idea to the public. Clearly, the techniques of risk

assessment must be brought into focus before the results will be useful. The application of such techniques to regulatory decision-making is a major public policy issue.

Attention must be called to the NPIRS system (3), the National Pesticide Information Retrieval System, which has been established to provide information on the registration of pesticides, specifically to the scientific, regulatory, and extension communities. NPIRS might be called upon by the news media to provide some of the background information which is presently lacking in the short time-lines which face reporters in collecting information on late-breaking stories. This is an on-line data base which requires only a telephone and a computer terminal to access and a little money for access charges. The system can provide the basic background information which should be translated into an understanding of the hazards associated with the use of pesticides.

Finally, although mass media rarely change existing attitudes, they do strengthen public perception. Scientists and journalists must work together to promote a better informed public sector. Accuracy, objectivity, and sourcing are identified as problems. The scientific community must endeavor to understand the media process and the constraints faced in covering scientific issues.

Commentary

Following the conclusion of the formal paper presentations in the Symposium on which this book is based, four individuals were asked to comment on the ideas and information presented from their own unique perspectives.

Anne E. Lindsay, Chief of the Policy and Special Projects Staff, Office of Pesticide Programs, U.S.E.P.A., speaking from the perspective of a regulatory official, was struck by the amount and complexity of information needed to answer what seem to be simple questions pertaining to whether we can use a product safely or whether food treated with pesticides is really safe to eat. As a regulatory agency, it is EPA's role to perform that task on the public's behalf: to take a lot of different kinds of information, evaluate it, and produce a very basic, public decision that a pesticide can or cannot be used.

The actual process of reaching such decisions is largely unknown outside EPA, and may seem to be a "black box." EPA engages in all three of the basic approaches to pesticide safety discussed in this book, namely hazard identification, exposure assessment and reduction, and communication efforts. Hazards are identified by requiring extensive testing before a pesticide is registered for use. One hundred or more individual tests for health and environmental effects are now required, depending on proposed uses. These requirements can be a "black box" for the industry, but EPA is trying to correct that impression by encouraging companies to consult with them early in the process to clarify any data development problems which might arise before making expensive testing commitments which might be inappropriate or unnecessary.

In the area of exposure, we are dealing with the variable on which we can really have an effect through regulation (unlike toxicity of a chemical). EPA is finding also in reviewing older pesticides that exposure was often not very carefully evaluated in the past, and applicators are frequently found at greater risk than previously thought. We can reduce the risk of exposure through changes in application rates and methods, geographic restrictions, protective clothing and other label precautions, and restricting use to trained applicators. The question of the adequacy of label precautions must be addressed.

Carol N. Scott, Executive Director of the Committee to Coordinate Environmental Health and Related Programs, U.S. Department of Health and Human Services, speaking from the perspective of the public policy analyst, noted that in an ideal political and social climate a risk assessment should, in advance of a crisis, present to knowledgeable risk managers a quantitative risk assessment which expresses all of the uncertainties incorporated in the assumptions. However, too often the crisis occurs first, and the public-press-politician synergistic relationship takes over, as in the case of the ethylene dibromide situation, or someone leaks misinformation to the press, as in the case of Alar, or politics takes over, as in the case of the controversy over the appropriate ways to use and regulate biotechnology.

The results of conducting the process of risk management in the public arena are all too often irrational decisions without the benefit of good scientific data. This can result in the commitment of resources and the expenditure of large sums of money to "control" emotionally volatile risks to the benefit of almost no one.

We cannot hope to change the press or the politicians, but we can educate the public. Risk managers should develop public policies in advance of the next crisis. No one in the public realm wishes to subject the public to unnecessary risk. But a small risk with a large benefit may be quite acceptable to the public, if the public is effectively presented with all the data. This emphasizes the need for effective public communication.

Dr. Donald D. Kaufman, Chief, Soil-Microbial Systems Laboratory, Agricultural Research Service, U.S.D.A., Beltsville, Maryland, expressed his belief that the bench scientist involved in pesticide research has in the past had too provincial a perspective with respect to the risks associated with pesticide use. Frequently these scientists are working with the chemicals early in the development process before introduction to the market. The bench scientist has a perspective on what the realistic risk of a future pesticide is and how that risk might be perceived. This perspective should be communicated and factored into the regulatory decision-making process. In the past this has not always been done effectively.

Have we asked the right questions? Have we asked the right questions at the right time? The pendulum seems to have swung back from past practice, and now many questions which should have been asked, are being asked. It is now acceptable for the researcher, the developer, the regulator, and the policy analyst to sit down together, consider the data, and make decisions. Rather than focusing on the negative, which is society's propensity, we need to take a balanced approach, using all the data available, both the positive

and negative aspects of a candidate pesticide's behavior, to develop
a realistic risk assessment.

We need to research how to use effectively our existing pesticides, how to conserve them, and how to fit them into new strategies
for pest control. In the past it was difficult to consider the use
of pheromones, other alternative controls, or biological controls
of pests because there was no funding incentive for development.
Now it is becoming a very realistic part of our research because it
has become apparent that there are fewer and fewer new pesticides
coming into the market. We must literally look at how we can get
basic principles back into the system in terms of control and save
some of the pesticide chemicals we have. We feel that we are in
danger of losing a major part of them, and we cannot afford that.
Wise use practices and intelligent assessment of the risks of pesticide use will permit us to prolong their life. Research is essential to achieve this.

Dr. James M. Witt, Extension Specialist in Chemistry and Toxicology, Department of Agricultural Chemistry, Oregon State University,
Corvallis, expressed his feeling that communication is a difficult
art. The final and most important aspect of the search for knowledge is its communication. This is especially so in the area of
risk from pesticides, where there are many simple assertions and
questions but no simple answers. The nature as well as use patterns
of pesticides and consequently their associated risk is governed by
how they are regulated. This in turn is governed by both the regulating agency's perceptions and the perceptions of the public. Dr.
Young identified an important reality when he quoted the ancient
maxim, "Perception is more important than reality."

The argument of risk from pesticides is expressed in the language of toxicology and chemistry, but the issue is often one of
philosophy. In our communication we must first teach the elements
of toxicology and chemistry necessary for understanding and interpretation of the reality of the data, provide an accurate summary
of the data, and attempt to isolate or separate the toxicological
arguments and risk evaluations from the philosophical and emotional
arguments. This is not easy; few audiences have either the time or
interest requisite to understanding the elements of risk evaluation.
We have consistently failed to teach the concept of dose/response -
that as you increase the dose, you increase the severity, frequency, and nature of the effects and as you decrease the dose, the
opposite happens.

We can identify pesticide risks and we can reduce them. But
can we adequately define to what level we wish to reduce them?
This is a philosophical as well as a social issue. At present there
may be no socially acceptable risk from chemicals similar to the
acceptable risks from food-borne infections, occupational injuries,
or injuries incurred around the home. The public often occupies an
extremist position - from "It can't hurt you; I've bathed in it,"
to "I don't care what you say, I don't want any exposure; I want
zero risk."

Society will decide what is a socially acceptable risk. Toxicologists should learn to communicate with the public to provide
the basis for such decisions. The chemical risks are summarized in

terms of margins of safety (for obvious chemical injury) and probability (for injury which might occur some time in the future). Until we can communicate the significance of a margin of safety of 10 versus 1,000, or a probability of 1×10^{-4} versus 1×10^{-9}, the social decisions on acceptable risks will be made from fear on the basis that any level of exposure is significant and all risks are equal.

Conclusion

This book, I believe, has identified many areas where we need substantially more data, highlighted some problem areas, and certainly given us new ways to look at some problems we have today. If the solution to a problem lies first in its identification, I believe we have taken a positive step forward.

Literature Cited

1. Ware, George W. "Pesticides: Theory and Application"; W. H. Freeman and Co.: San Francisco, 1983; p. 5-6.
2. National Research Council. "Pesticide Resistance: Strategies and Tactics for Management"; National Academy Press: Washington, D. C., 1986; 471 pp.
3. National Pesticide Information Retrieval System, an on-line data base managed by Purdue University, West Lafayette, Indiana.

RECEIVED December 11, 1986

Author Index

Cardona, Raymond A., 14
de Serres, Frederick J., 37
Dorough, H. Wyman, 106
Hart, Ronald W., 141
Hock, W. K., 130
Hollingworth, Robert M., 54
Kaufman, Donald D., 174
Lindsay, Anne E., 173
Madden, L. V., 77
Menzer, Robert E., 170

Scott, Carol N., 174
Seiber, James N., 88
Stevens, J. T., 43
Sumner, D. D., 43
Turturro, Angelo, 141
Valenti, JoAnn Myer, 151
Vorpagel, Erich R., 115
Witt, James M., 175
Yang, Raymond S. H., 20
Young, Alvin L., 1

Affiliation Index

Ciba-Geigy Corporation, 43
E. I. du Pont de Nemours & Co., 115
Mississippi State University, 106
National Center for Toxicological Research, 141
National Institute of Environmental Health Sciences, 20,37
Office of Science and Technology Policy, 1
Ohio State University, 77
Oregon State University, 175

Pennsylvania State University, 130
Purdue University, 54
U.S. Department of Agriculture, 174
U.S. Department of Health and Human Services, 174
U.S. Environmental Protection Agency, 173
Uniroyal Chemical Company, 14
University of California, 88
University of Maryland, 170
University of Tampa, 151

Subject Index

A

Above-ground soil disposal beds, pesticide disposal, 139-141
Acceptable daily intake (ADI), pesticides, 17-18
Acetohydroxyacid synthase, inhibition in plants, 66
Acetolactate synthase
  catalysis mechanism, 116-118
  inhibition, 121
  inhibitors, 118
  purpose, 116
2-Acetylaminofluorine
  carcinogenicity, 108
  use in evaluation of carcinogenicity, 39-40
Acetylcholinesterase, comparative inhibition, 61f,110

Acrylonitrile, 38
Acute, definition, 132
Acute toxicity, 2,4-D, 24
  definition, 20
  determination, 14-15
  focus, 172
  hexachlorobenzene, 24-25
  organophosphates, 23
  pyrethroids, 26
  testing purpose, 15
  tissues involved, 23
  See also Toxicity
ADAPT
  model function, 46t
  pattern recognition modeling, 46
Advanced Continuous Simulation Language (ACSL), application, 30
Agenda setting, mass media, 158
Aggregation, pests, 80

Agricultural workers, risks of
    pesticides, 133
Agrochemicals
    application technology, 135-136
    discovery, 115
    See also Pesticides
Air-assist spraying, description, 136
Aldicarb, use on Long Island, 174
Americans, voluntary risks, 147
Ames test
    complementary tests, 38-39
    description, 37
    use of 2-acetylaminofluorine, 108
Apple growers, insecticide use, 80
Aryloxadiazolone anticholinesterases,
    inhibition in resistant
        strains, 62
Asipu, risk assessors, 143-144
Atmosphere, pesticide entry, 98
Atrazine, rainwater concentration, 101
Attitude of acceptance, 157
Avermectins, toxicity, 55,56

                            B

Bacillus thuringiensis, 57,71
Benomyl, impact on the environment, 6
Benzene, 38
Benzimidazole fungicides, binding
    site, 68
Benzoin, 38
Bhopal, media coverage, 154-155
Bioaccumulation, definition, 94
Bioconcentration factor, description
    and use, 94-95
Biodegradation, estimation
    techniques, 96
Bioevaluation, improvement, 63-65
Biomagnification, definition, 94
Biorational design
    sulfonylureas, 119-127
    target identification, 119-121
    use, 115

                            C

California, groundwater contamination
    act, 97-98
Canavanine
    structure, 72f
    toxicity, 71
Caprolactam, 38
Carbamate ester hydrolysis, rats, 112t
Carbamate insecticides
    hydrolysis in rats, 112
    mode of action, 110

Carbaryl, toxicity, 110
Carbon tetrachloride, enhancement of
    acute toxicity by kepone, 28t
Carcinogenicity
    chemicals evaluated as positive in
        humans, 48t
    evaluation scheme, 40-42
    malathion, 23
    methyl parathion, 23-24
    pyrethroids, 27
    tetrachlorvinphos, 24
    trichlorfon, 23
Carcinogens, genotoxic and epigenetic
    effects, 18
Chemical structure
    activity relations, 8
    relationship to toxicological
        response, 43-47
Chemoreception, use, 57
Chlordecone
    effects on toxicity of carbon
        tetrachloride, 27
    injuries to workers, 55
Chlorobenzene, polarity, 91
Chlorpyrifos, fog water
    concentration, 101
Chlorsulfuron
    metabolized form, 118
    selectivity as a herbicide, 118
Chronic effects, definition, 132
Chronic exposure, pharmacokinetics, 21
Chronic high, definition, 132
Chronic low, definition, 132
Chronic tests, requirements, 16
Chronic toxicity
    comparison with acute toxicity, 21t
    definition, 20
    1,3-dichloropropene, 25
    focus, 172
    gray areas, 24
    nontarget species, 72
    organophosphates, 23-24
    pyrethroids, 26-27
    self-propagating effects, 23
    See also Toxicity
Cockroach, toxic response, 64f
Computer-assisted synthesis planning
    (CASP), model function, 46t
Computers, use in biorational design
    of pesticides, 115-116
Constrained structure generation
    (CONGEN), model function, 46t
Cooperative Extension Service, 174
Cotton, insecticide use, 79
Crop protection chemicals
    criteria, 127
    molecular modeling use in
        design, 121-122
Crop selectivity, acetolactate
    synthase, 118
Crops, pest density relationship, 84

Cross-resistance
  negatively correlated, 60
  See also Resistance

D

DDT
  bird population effect, 55
  rainwater concentration, 101
Department of Agriculture, pesticide
    emphasis, 73
Department of Health and Human
    Services, report on risk
    assessment, 144
Dermal penetration study, requirements
    needed for registration, 17
Diazinon, 101
Dibromochloropropane, injuries to
    workers, 55
2,4-Dichlorophenoxyacetic acid
    (2,4-D), acute and chronic
    toxicity, 25-26
1,3-Dichloropropene
  acute and chronic toxicity, 25-26
  use, 26
Dichlorvos, neurotoxicity, 24-25
Dielectric constant,
    measurement, 89-91
Diethylhexylphthalate, 38
Diethylstilbestrol, 38
Dipole moment, measurement, 89-91
Disulfiram
  influence on the chronic toxicity of
      ethylene dibromide, 27-28
  interaction with ethylene
      dibromide, 28t
Dose-response assessment,
    definition, 145
Dose-response relationship,
    concept, 177
DRASTIC, groundwater contamination
    potential, 97
Drift, pesticide sprays, 136

E

Ecological magnification,
    definition, 94
Ecosystem
  proper use of pesticides, 84
  role of the pest, 77
Electrostatic sprayers
  applicator exposure, 137t
  description, 136
Environmental chemodynamics,
    definition, 103

Environmental fate
  concerns, 88
  determination using EXAMS, 99
  minimizing risks, 88-103
  physicochemical properties
      involved, 89
Environmental Protection Agency (EPA)
  approaches to pesticide safety, 175
  role in FIFRA, 14
  toxicity studies, 14
  toxicology data requirements, 15,172
Environmental risk
  reporting, 156-157
  See also Risk
Environmentalists, opinions on risk
    coverage, 161
Enzymes
  assays, 65
  characterization, 121
  inhibitor design, 121-127
Esterase inhibitors, organophosphate
    resistance prevention, 60
Ethylbenzimidazole, fog water
    concentration, 101
Ethylene dibromide
  major histopathological findings in
      rats, 29t
  media coverage, 162-163
EXAMS, tool for estimating
    volatilization, 101
Exposure assessment
  definition, 145
  modeling system, 99

F

Federal Insecticide, Fungicide, and
    Rodenticide Act (FIFRA), role of
    the EPA, 14
Flux values, chemicals in flooded rice
    fields, 100t
Fog water
  distribution of chemicals with
      interstitial air, 102t
  pesticides, 101-103
Food and Drug Administration
  degree of testing required, 45
  presumptive toxicity
      classifications, 45t
Fragment constant approach,
    calculation of octanol-water
    partition coefficient, 93-94
Fungicides
  use in North Carolina, 80
  See also Pesticides

G

Genotoxicity tests, new
   approaches, 37-42
Genotoxins, in vitro, 40
Goggles, protective items for
   pesticide handlers, 134
Groundwater
   measurement of contamination
      potential, 97
   pesticide contamination, 6,107
Guinea pigs, dermal sensitization, 15

H

Halogenated hydrocarbons, acute and
   chronic toxicity, 25-26
Hansch regression equation, 44
Health, role in risk decision, 151
Heliothis, pesticide efficiency, 57
Henry's law constant, use, 95
Heptachlor, toxicity, 109-110
Heptachlor epoxide, toxicity, 109-110
Herbicides
   long-term consequences, 81
   photosynthesis inhibition, 66
   U.S. sales, 78
   use by U.S. farmers, 3
   See also Pesticides
Hexachlorobenzene
   porphyria-induced, 26
   use, 25
Hexamethylphosphoramide, 38
Host resistance, 81
Hydrolysis, estimation techniques, 96

I

Imidazolinones, enzyme inhibition, 66
Immunotoxicology, research needs, 18
In vitro assay
   sulfonylurea activity, 117
   use in pesticide development, 65
   See also Short-term in vitro tests
In vivo tests
   role of hydrolysis, 112
   See also Short-term in vivo tests
Inhibitors, design using molecular
   modeling, 127
Insecticides
   degradation in mammals, 71
   risk to humans, 56
   use by U.S. farmers, 3
   yield benefits, 4
   See also Pesticides

Insects, defense response against
   pathogens, 71
Integrated pest management,
   description, 80
International Agency for Research on
   Cancer, 23
International Program on Chemical
   Safety (IPCS), 38
Isobutylamides, housefly toxicity, 62

J

Johnsongrass, location, 79
Journalists
   improving communication with
      scientists, 163
   precision, 155-156
   view versus scientist's
      view, 159-160

L

Late blight, location, 79
Leaching, current understanding, 103
Leptophos, neurotoxicity, 24-25
Linear free energy relationships,
   estimation of rate constants, 95
London forces, 91

M

Malaria, effect of DDT, 5
Malathion
   carcinogenic potential, 23-24
   rainwater concentration, 101
Malaxon, carcinogenicity, 24
Mammalian metabolism
   minimizing pesticide risk, 106-114
   relationship to pesticide
      risk, 107-109
   understanding versus
      knowledge, 107-108
Manduca larvae, production of
   antibiotic peptides, 72
Mass media
   effect on public perceptions of
      pesticide risk, 153-168
   flow in science coverage, 155
   process, 161-162
   relationship with science, 156
   role in reporting risks, 157-159,175
   science literacy, 159
   science sources, 153-168

Mass media--Continued
  type of science covered, 160
  view of public, 154-155
  view of scientists, 154
Maximum tolerated dose (MTD), chronic
    toxicity studies, 17
Media coverage, recommendations, 167
Median lethal dose, definition, 15
Merphos, neurotoxicity, 24-25
Metabolic processes
  pesticide persistence, effect, 111
  relationship to toxicity, 111-113
Methyl parathion
  carcinogenic potential, 23-24
  partition coefficient value, 93
The Miami Herald, ethylene
    dibromide story, 163
Mirex, minimizing exposure, 56
Mitochondrial poisons, use, 60
Models, use in environmental fate
    studies, 89
Molecular descriptors, used in
    structure-activity
    relationships, 46t
Molecular modeling
  designing crop protection
      chemicals, 115-127
  use, 122
Molinate
  flux calculation with EXAMS, 100
  toxicity to carp, 57
  volatilization rate, 100
Monooxygenases, potential target
    site, 68
Mortality, rats in ethylene
    dibromide-sulfiram interaction
    study, 29t
Mouse bone marrow micronucleus test,
    in vitro, 40
Mutagenicity studies, requirements
    needed for registration, 17

N

1-Naphthol, toxicity, 110
National Institute of Occupational
    Safety and Health, 28
National Pesticide Information
    Retrieval System (NPIRS)
  description, 8
  purpose, 175
National Toxicology Program, 24
New Jersey, environmental risk
    reporting, 156-157
Nitrobenzene, polarity, 91
Nitromethylene heterocyclic
    insecticides, housefly
    toxicity, 63

p-Nitrophenol, fog water
    concentration, 101
N-Nitrosamine groups, association with
    animal tumors, 44
Nitroso compounds,
    carcinogenicity, 108
No observed effect level (NOEL)
  definition, 18
  description, 45
  purpose, 17

O

Octanol-water partition coefficient
  description, 92-94
  estimation of bioconcentration
      factor, 95
  See also Partition coefficient
Oncogenicity studies, purpose, 16
Organic chemicals, bioconcentration
    factor range, 94
Organophosphate poisoning, humans, 23
Organophosphates
  acute and chronic toxicity, 23-25
  anticholinesterase property, 23
  myopathy, 25
  neurotoxicity, 24

P

Paraoxon
  fog water concentration, 101
  solubility, 92
  toxicity, 110
Paraquat, hazard, 58
Parathion
  carcinogenic potential, 23-24
  fog water concentration, 101
  rainwater concentration, 101
  toxicity, 110
Partition coefficient
  measurement, 93
  See also Octanol-water partition
      coefficient
Pattern recognition modeling,
    ADAPT, 46
Personal protective equipment,
    pesticides, 134-135
Pest density
  crop yield relationship, 84
  threshold level, 83
  variation, 79
Pest resistance, pesticides, 173
Pest vulnerability, discovery, 68-72
Pesticide metabolites
  relationship to in vivo
      toxicity, 107

Pesticide metabolites--Continued
  toxicity, 109-110
  toxicology testing, 107-108
Pesticide registration
  balanced approach, 176-177
  role of scientific review
    panel, 113-114
  toxicology requirements, 14-19
Pesticide research
  federal and state expenditures, 9t
  funding, 9
  industrial expenditures, 9t
Pesticide residues
  development of a national policy, 8
  in food, 55
Pesticide risks, knowledge used on new
    materials, 109
Pesticides
  acute and chronic toxicity, 23-27
  air-water distributions, 98-103
  applicators, measuring exposure, 141
  benefits, 4-5
  chronic adverse effects, 133
  controversy, 54
  cost of development, 62
  defining adequate levels, 177
  definition, 3
  disposal, innovative technology, 141
  educational task of industry, 158
  effective use, 177
  EPA's role in safety, 175-176
  focus of controversy, 1
  fog water, 101-103
  future, 72-74
  hazards, 174
  health threats, 132
  human fatalities, 54-55
  identifying metabolites, 107
  in the United States, 133
  interactions in the environment, 173
  media coverage, 154,161
  metabolic processes and
    toxicity, 111-113,174
  minimizing exposure, 56-57
  minimizing risk, 172-178
  necessity, 55
  need for proper protection, 132-141
  new and safer, 62-72
  personal protective
    equipment, 134-135
  pest resistance, 173
  populations at risk, 133
  potency of new compounds, 55
  public information, 156
  regulation based on physicochemical
    properties, 103
  resistance, strategies to
    overcome, 58-62
  rinse water, recycling, 139
  risk
    communication, 177
    description, 6-7,56

Pesticides--Continued
  minimization, 106-114
  reporting, 161
  safety record, 30
  sources for media coverage, 156
  spray tank transfer, 139
  sprayable formulations, 136
  synergism, 57-58
  systemic poisoning, 133
  topical effects, 133
  toxicity modification, 57-58
  toxicology
    federal expenditures for
      research, 10t
    interactions, 27
  use
    in the United States, 2t,78
    on major food crops, 79t
    minimizing the risk, 1-10
    volatilization from flooded
      fields, 98
  waste disposal, 138-141
  world market, 3
  See also Agrochemicals, Fungicides,
    Herbicides, Insecticides
Pests
  aggregation, 80
  associated produce losses, 78t
  classification, 83
  definition, 77
  effect on crop yields, 83-84
  impact, 77-78
  interaction with the
    environment, 81,173
  modes of action, 66
  part of the ecosystem, 77-84
  population dynamics, 81-83
  response-stimulus relationships, 81
  targeting pesticides, 6
  vulnerability, 54-73
  worldwide food losses due to, 4
Pharmacokinetics
  chemicals between two species, 30
  physiologically based, 30
Phenobarbitone, 38
N-Phenylcarbamates, fungi toxicity, 62
Pheromones, use, 56,177
Philippines, pesticide-related
    deaths, 55
Phorate sulfoxide, solubility, 92
Photolysis, estimation techniques, 96
N-Phthalyl-L-valine anilide, Zea mays
    toxicity, 67f
Physicochemical properties
  environmental relevance, 96-98
  measurement, 89
Plague, effect of DDT, 5
Polarity
  definition, 89
  role in environmental fate, 89-91
Potatoes, diseases, 79
Predictive toxicology, principles, 47

Produce losses, 78
Professional societies, aid in risk
    management decisions, 150
N-Propylcarbamates, leafhopper
    toxicity, 60-62
Public, role in risk assessment, 151
Public education
    toxicity risks, 143-152
    understanding risk, 147-148
Public perceptions
    conclusions, 147
    issues of belief, 157
Pyrethroids
    acute and chronic toxicity, 26-27
    resistance mechanism, 59
    toxicity, 55
    toxicity site, 56
    See also Synthetic pyrethroid
        insecticides
Pyruvate, condensation with thiamin
    pyrophosphate, 117

Q

Quantitative structure-activity
    relationships (QSAR)--See
    Structure-activity relationships

R

Rainwater, chemicals contained in, 101
Rate constants
    availability of data, 96t
    environmental fate use, 95
Research
    benefits, 84
    pesticide interactions in the
        environment, 173
    pesticide-related, 73
    public's use of mass media, 156
    risk management, 152
Resistance
    alleviation strategies, 59-60
    mechanisms for monitoring, 59
    origin, 59
    problem, 7
    tests for enzymological markers, 59
    See also Cross-resistance
Resistance management, critical
    areas, 58-59
Respirator, protective items for
    pesticide handlers, 134
Rice, cause of crop loss, 78
Risk
    acceptability, 145-146
    assumptions, 146
    considerations, 146-147

Risk--Continued
    definition, 145
    estimation, 146-147
    factors in understanding, 145-148
    minimization
        via chemical
            understanding, 173-174
        via metabolism studies, 113-114
        via pest studies, 173
        via pesticide hazard
            understanding, 174-175
    perception, 147
    pesticides, 6-7,56
    population, 146
    prediction
        by animal experimentation, 18
        by metabolite knowledge, 108-109
    public education, 143-152
    reduction by technological
        means, 150
    research approaches to minimize, 7-9
    uncontrollable, 146
    understanding, 147-148
    voluntary aspects, 147
    See also Environmental risk
Risk assessment
    analysis, 149
    basis, 149
    comparison importance, 149
    definition, 145
    problem-solving approach, 144
    purpose, 176
    responsible media coverage, 163-168
Risk characterization, definition, 145
Risk communication
    media role, 166
    pesticides, 177
    useful guidelines, 165
Risk management
    definition, 145
    problem-solving approach, 148-151
    research, 150
    results in the public arena, 176
Risk management decision
    analysis of possible
        options, 149-151
    basis, 149
    four sections, 148
    evaluation of options, 151
    information dissemination, 151
    promotion of acceptance, 151
    range of options to be
        considered, 150

S

Safety, environmental tests, 88
Safrole, 38
Salmonella, use in carcinogenic
    tests, 37,39

Science news, popularity, 165-166
Science reporting, science
    literacy, 159
Scientific review, pesticides, 113-114
Scientists
    media preference, 166
    respect for reporters, 154
Sesamex, synergism, 63
Short-term in vitro tests
    effective deployment, 40-42
    evaluation, 38-39
    proposed testing scheme, 41f
    See also In vitro assay
Short-term in vivo tests
    effective deployment, 40-42
    evaluation, 39
    proposed testing scheme, 41f
    See also In vivo tests
Simazine, rainwater concentration, 101
Simulation modeling, toxicology, 43-48
Soil
    chemical mobility, 96
    leaching index, 97
Soil degradation, sulfonylureas, 119
Soil mobility, relationship to water
    solubility, 97t
Sri Lanka, pesticide-related
    deaths, 55
Structure-activity
    relationships, 43-48
Subchronic tests, duration, 15-16
Suicide substrates, design, 121
Sulfonylureas
    activity in in vitro assay, 117-118
    biorational design, 119-127
    biosynthesis inhibition, 119
    description, 116
    enzyme inhibition, 66
    soil degradation, 119,120f
Supercooled liquid, water
    solubility, 92
Synthetic pyrethroid insecticides
    mode of action, 109
    See also Pyrethroids

T

Tampa Tribune, ethylene dibromide
    story, 162
Television, science coverage, 160
Teratogenesis, research needs, 18
Teratogenicity, effect on
    reproduction, 16
Tetrachlorvinphos, 24
Thiobencarb, volatilization rate, 100
Thiocarbamates, 57,92
Thiophosphates
    conversion to phosphates, 111-112
    inhibitors of
        acetylcholinesterase, 112

Thyroxin, computer representation, 122
Thyroxin-prealbumin binding
    cross section of the pest and
        nonpest binding sites, 125f-126f
    surface, 123f
Toluene, polarity, 91
O-Toluidine, 38
Toxicity
    avermectins, 55
    canavanine, 71
    carbaryl, 110
    modification, 57-58
    paraoxon, 110
    parathion, 110
    pesticide metabolites, 109-110
    pyrethroids, 55
    relationship to metabolic
        processes, 111-113
    See also Acute toxicity, Chronic
        toxicity
Toxicity data, extrapolation, 32
Toxicological interactions,
    evaluation, 30
Toxicology
    computer-assisted research
        models, 46t
    minimizing risk by
        understanding, 172-173
    pesticides, 13
    research needs, 7-8
    simulation modeling, 43-48
Trichlorfon
    carcinogenicity, 24
    neurotoxicity, 24-25
Tumor, increase after disulfiram-
    ethylene dibromide exposure, 29
Typhus, effect of DDT, 5

V

Vegetable oils, carrier use in
    pesticide applications, 138
Volatilization
    estimation techniques, 96
    importance as a fate process, 100
    measurement, 95,98-99

W

Waste disposal
    pesticides, 138-141
    sources of potential problems, 139
Water solubility
    criterion for soil movement, 97
    description, 91-92
    determination of octanol-water
        partition coefficient, 93

Waterproof items, for pesticide
    handlers, 134
Weibull model, 18
Wheat, cause of crop loss, 78

                    X

Xenobiotics, performance of
    mechanistic studies, 43

                                Y

            Yellow fever, effect of DDT, 5

Production by Paula M. Bérard
Indexing by Keith B. Belton
Jacket design by Amy Hayes

Elements typeset by Hot Type Ltd., Washington, DC
Printed and bound by Maple Press Co., York, PA

Recent ACS Books